Spatio-Temporal Modeling of Nonlinear Distributed Parameter Systems

International Series on
INTELLIGENT SYSTEMS, CONTROL, AND AUTOMATION:
SCIENCE AND ENGINEERING

VOLUME 50

For other titles published in this series, go to
www.springer.com/series/6259

Han-Xiong Li • Chenkun Qi

Spatio-Temporal Modeling of Nonlinear Distributed Parameter Systems

A Time/Space Separation Based Approach

Han-Xiong Li
City University of Hong Kong
Dept of Manufacturing
Engineering and
Engineering Management
Hong Kong
China, People's Republic

and

Central South University
School of Mechanical and
Electrical Engineering
Changsha
China, People's Republic
E-mail: mehxli@cityu.edu.hk

Chenkun Qi
Shanghai Jiao Tong University
School of Mechanical Engineering
Shanghai
China, People's Republic
E-mail: chenkqi@sjtu.edu.cn

ISBN 978-94-007-0740-5 e-ISBN 978-94-007-0741-2
DOI 10.1007/978-94-007-0741-2
Springer Dordrecht Heidelberg London New York

Typesetting & Cover design: Scientific Publishing Services Pvt. Ltd., Chennai, India

Printed on acid-free paper

Springer is part of Springer Science+Business Media (www.springer.com)

Preface

Distributed parameter systems (DPS) widely exist in many industrial processes, e.g., thermal process, fluid process and transport-reaction process. These processes are described in partial differential equations (PDE), and possess complex spatio-temporal coupled, infinite-dimensional and nonlinear dynamics. Modeling of DPS is essential for process control, prediction and analysis. Due to its infinite-dimensionality, the model of PDE can not be directly used for implementations. In fact, the approximate models in finite-dimension are often required for applications. When the PDEs are known, the modeling actually becomes a *model reduction* problem. However, there are often some unknown uncertainties (e.g., unknown parameters, nonlinearity and model structures) due to incomplete process knowledge. Thus the *data-based modeling* (i.e. *system identification*) is necessary to estimate the models from the process data. The model identification of DPS is an important area in the field of system identification. However, compared with traditional lumped parameter systems (LPS), the system identification of DPS is more complicated and difficult. In the last few decades, there are many studies on the system identification of DPS. The purpose of this book is to provide a brief review of the previous work on model reduction and identification of DPS, and develop new spatio-temporal models and their relevant identification approaches. All these work will be presented in a unified view from time/space separation. The book also illustrates their applications to thermal processes in the electronics packaging and chemical industry.

In the book, a systematic overview and classification on the modeling of DPS is presented first, which includes model reduction, parameter estimation and system identification. Next, a class of block-oriented nonlinear systems in traditional LPS is extended to DPS, which results in the spatio-temporal Wiener and Hammerstein systems and their identification methods. Then, the traditional Volterra model is extended to DPS, which results in the spatio-temporal Volterra model and its identification algorithm. All these methods are based on linear time/space separation. Sometimes, the nonlinear time/space separation can play a better role in modeling of very complex process. Thus, a nonlinear time/space separation based neural modeling is also presented for a class of DPS with more complicated dynamics. Finally, all these modeling approaches are successfully applied to industrial thermal processes, including a catalytic rod, a packed-bed reactor and a snap curing oven.

The book assumes a basic knowledge about distributed parameter systems, system modeling and identification. It is intended for researchers, graduate students and engineers interested in distributed parameter systems, nonlinear systems, and process modeling and control.

Authors are grateful to students, colleagues and visitors in our research group for their support and contributions, and also would like to thank the Research Grant Council of Hong Kong and National Natural Science Foundation of China for their financial support to our research. Last, but not least, we would like to express our deepest gratitude to our wives, children and parents for their love, understanding and support.

Han-Xiong Li
City University of Hong Kong
Central South University
Chenkun Qi
Shanghai Jiao Tong University

Contents

List of Figures

List of Tables

Abbreviations

AE	Algebraic Equation
AIM	Approximated Inertial Manifold
BF	Basis Function
DE	Difference Equation
DPS	Distributed Parameter System
EEF	Empirical Eigenfunction
EF	Eigenfunction
ERR	Error Reduction Ratio
FDM	Finite Difference Method
FEM	Finite Element Method
FMNS	Fading Memory Nonlinear System
IC	Integrated Circuit
IM	Inertial Manifold
IV	Instrumental Variables Method
KL	Karhunen-Loève Decomposition
KL-Hammerstein	Karhunen-Loève based Hammerstein Model
KL-Wiener	Karhunen-Loève based Wiener Model
LDS	Lattice Dynamical System
LPS	Lumped Parameter System
LSE	Least-Squares Estimation
LTI	Linear Time Invariant
MARE	Mean of Absolute Relative Error
MIMO	Multi-Input-Multi-Output
MO	Multi-Output
MOL	Method of Lines

NARX	Nonlinear Autoregressive with Exogenous Input
NL-PCA	Nonlinear PCA
NL-PCA-RBF	NL-PCA based RBF model
ODE	Ordinary Differential Equation
OFR	Orthogonal Forward Regression
PCA	Principal Component Analysis
PCA-RBF	PCA based RBF model
PDE	Partial Differential Equation
POD	Proper Orthogonal Decomposition
RBF	Radial Basis Function
RMSE	Root of Mean Squared Error
SISO	Single-Input-Single-Output
SNAE	Spatial Normalized Absolute Error
SNR	Signal-to-Noise Ratio
SO	Single-Output
SP-Hammerstein	Spline Functions based Hammerstein Model
SP-Wiener	Spline Functions based Wiener Model
SVD	Singular Value Decomposition
TNAE	Temporal Normalized Absolute Error
WRM	Weighted Residual Method

1 Introduction

Abstract. This chapter is an introduction of the book. Starting from typical examples of distributed parameter systems (DPS) encountered in the real-world, it briefly introduces the background and the motivation of the research, and finally the contributions and organization of the book.

1.1 Background

Advanced technological needs, such as, semiconductor manufacturing, nanotechnology, biotechnology, material engineering and chemical engineering, have motivated control of material microstructure, fluid flows, spatial profiles (e.g., temperature field) and product size distributions (Christofides, 2001a). These physical, chemical or biological processes all lead to so called distributed parameter systems (DPS) because their inputs and outputs vary both temporally and spatially. As the significant progress in the sensor, actuator and computing technology, the studies of distributed parameter processes become more and more active and practical in science and engineering. Recently several special issues for control of DPS have been organized by Dochain *et al.* (2003), Christofides (2002, 2004), Christofides & Armaou (2005), and Christofides & Wang (2008). Modeling is the first step for many applications such as prediction, control and optimization. This book will focus on the modeling problem of nonlinear DPS with application examples chosen as industrial thermal processes. In general, the modeling approaches presented are applicable to a wide range of distributed parameter processes.

Next, we will introduce some typical thermal process in integrated circuit (IC) packaging and chemical industry, which will be used as examples in the rest of chapters.

1.1.1 Examples of Distributed Parameter Processes

a) Thermal Process in IC Packaging Industry

One important thermal process in the semiconductor back-end packaging industry considered in this book is the curing process (Deng, Li & Chen, 2005). After the required amount of epoxy is dispensed on the leadframe from the dispenser, and a die is moved from the wafer to attach on the leadframe by the bond arm, then the bonded leadframe is moved into the snap curing oven to cure at a specified temperature. As shown in Figure 1.1, the snap curing oven is an important equipment to provide the required curing temperature distribution. The oven has four heaters for heating and four thermocouples for temperature sensing in the operation. The parts to be cured will be moved in and out from inlet and outlet, respectively.

H.-X. Li and C. Qi: Spatio-Temporal Modeling of Nonlinear DPS, ISCA 50, pp. 1–12.
springerlink.com

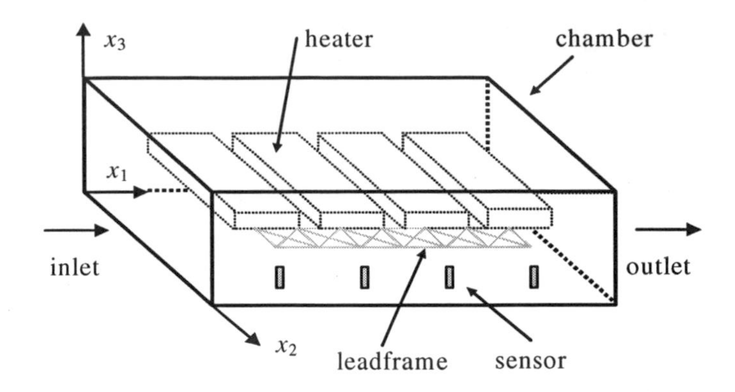

Fig. 1.1 Snap curing oven system

The temperature distribution inside the chamber is often needed for a quality curing control as well as the fundamental analysis for oven design. In practice, it is difficult to place many sensors to measure the temperature distribution during the curing. Thus it motivates us to build a model of the oven and use it to estimate the temperature distribution of the curing process.

This thermal process can be simplified for the easy modeling. The volume of the epoxy between a die and the leadframe is much smaller as compared to the volume of the leadframe and the volume of the oven chamber. Also, the volume of the leadframe is much smaller as compared to the volume of the oven chamber. Thus, the effects of the epoxy and the leadframe on the temperature in the oven chamber are usually neglected in modeling of the curing process. These effects can be considered as disturbances and could be compensated in the later control process.

This thermal process will follow the basic principles of the heat transfer (conduction, radiation and convection). The fundamental heat transfer equation of the oven can be expressed as a nonlinear parabolic partial differential equation (PDE) with some unknown parameters, unknown nonlinearities and unknown boundary conditions:

$$\rho(T)c\frac{\partial T}{\partial t} = \frac{\partial}{\partial x_1}\left(k(T)\frac{\partial T}{\partial x_1}\right) + \frac{\partial}{\partial x_2}\left(k(T)\frac{\partial T}{\partial x_2}\right) + \frac{\partial}{\partial x_3}\left(k(T)\frac{\partial T}{\partial x_3}\right) + f_c(T) + f_r(T) + bu(t), \tag{1.1}$$

where

$T = T(x_1, x_2, x_3, t)$ is the temperature at time t and location (x_1, x_2, x_3),

$x_1 \in [0, x_{10}]$, $x_2 \in [0, x_{20}]$ and $x_3 \in [0, x_{30}]$ are spatial coordinates,

$k(T)$ is the thermal conductivity, which is usually inaccurately given,

$\rho(T)$ is the density, which is usually inaccurately given,

c is the specific heat,

$f_c(T)$ and $f_r(T)$ are the effects of convection and radiation respectively, which are usually inaccurately given,

$u(t) = [u_1(t), u_2(t), u_3(t), u_4(t)]^T$ denotes the vector of manipulated inputs with the spatial distribution $b = [b_1(x_1, x_2, x_3), b_2(\cdot), b_3(\cdot), b_4(\cdot)]$,

$T_0 = T(x_1, x_2, x_3, 0)$ is the initial condition.

The boundary conditions are nonlinear functions of the boundary temperature, the space coordinates of the oven (x_1, x_2, x_3) and the ambient temperature T_a as follows:

$$k\frac{\partial T}{\partial x_i}\bigg|_{x_i=0} = f_{i1}(x_1, x_2, x_3, T, T_a)|_{x_i=0},$$

$$k\frac{\partial T}{\partial x_i}\bigg|_{x_i=x_{i0}} = f_{i2}(x_1, x_2, x_3, T, T_a)|_{x_i=x_{i0}}, i = 1,2,3,$$

However, it is difficult to obtain these nonlinear functions f_{i1} - f_{i2} using only physical insights.

b) Thermal Process in Chemical Industry

Catalytic Rod

Consider a long thin rod in a reactor as shown in Figure 1.2, which is a typical thermal process in chemical industry (Christofides, 2001b). The reactor is fed with pure species A and a zero-th order exothermic catalytic reaction of the form $A \rightarrow B$ takes place in the rod. Since the reaction is exothermic, a cooling medium that is in contact with the rod is used for cooling.

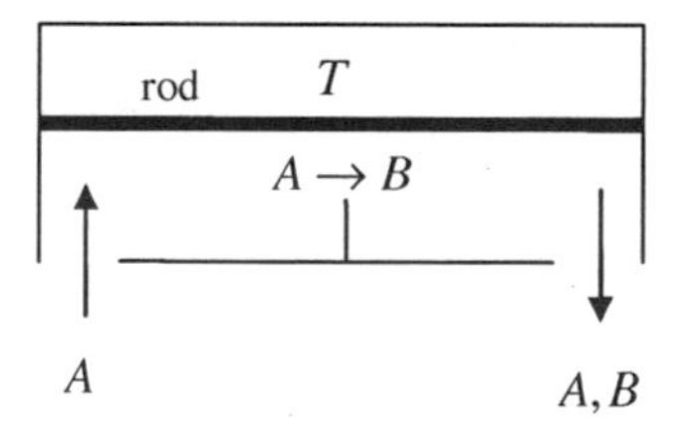

Fig. 1.2 A catalytic rod

Under the assumptions of constant density and heat capacity, constant conductivity of the rod, constant temperature at both sides of the rod, and excess of species A in the furnace, the nominal mathematical model which describes the spatio-temporal evolution of the rod temperature consists of the following parabolic PDE (Christofides, 2001b):

$$\frac{\partial T}{\partial t} = \frac{\partial^2 T}{\partial x^2} + f(T, \theta) + b(x)u(t), \tag{1.2}$$

subject to the Dirichlet boundary conditions:

$$T(0,t) = 0, \ T(\pi,t) = 0,$$

and initial conditions:

$$T(x,0) = 0,$$

where T denotes the temperature in the reactor, $u(t)$ denotes the manipulated input (temperature of the cooling medium) with the actuator distribution $b(x)$. θ denotes some unknown system parameters (e.g., heat of reaction, heat transfer coefficient and activation energy). f is an unknown nonlinear function.

Packed-Bed Reactor

Consider the temperature distribution in a long, thin non-isothermal catalytic packed-bed reactor in chemical industry (Christofides, 1998). As shown in Figure 1.3, a reaction of the form $A \rightarrow B$ takes place on the catalyst. The reaction is endothermic and a jacket is used to heat the reactor.

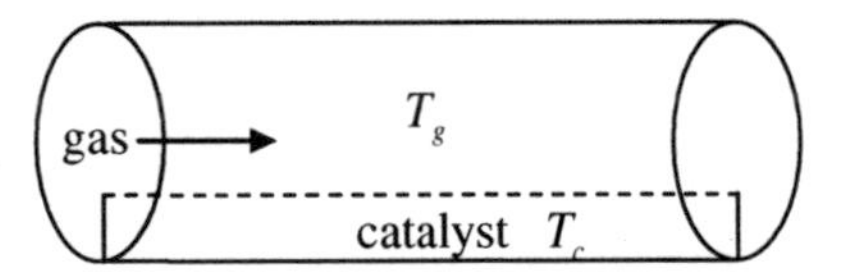

Fig. 1.3 A catalytic packed-bed reactor

Under the assumptions of negligible diffusive phenomena for the gas phase, constant density and heat capacity of the catalyst and the gas, and excess of species A in the reactor, a nominal dimensionless model that describes the temperature dynamics of this nonlinear tubular chemical reactor is provided as follows (Christofides, 1998)

$$\frac{\partial T_g}{\partial t} = -\frac{\partial T_g}{\partial x} + f(T_g, T_c, \theta) + b(x)u(t), \tag{1.3}$$

$$\frac{\partial T_c}{\partial t} = \frac{\partial^2 T_c}{\partial x^2} + g(T_g, T_c, \theta) + b(x)u(t), \tag{1.4}$$

subject to the boundary conditions

$$x = 0, \frac{\partial T_c}{\partial x} = 0, T_g = 0,$$
$$x = 1, \frac{\partial T_c}{\partial x} = 0,$$

where T_g and T_c denote the temperature of the gas and the catalyst, respectively. u denotes the manipulated input (the temperature of the jacket) with the actuator distribution $b(x)$. θ are unknown system parameters. f and g are unknown nonlinear functions.

1.1.2 Motivation

The industrial thermal processes mentioned are typical nonlinear distributed parameter systems that have following common features:

- These processes are usually described in partial differential equations (PDE) that show time/space coupled dynamics.
- The dynamics of these processes are strongly nonlinear.
- There exist uncertainties (e.g., unknown structure, unknown parameters, and external disturbances, etc.) in the process.

These features are also common to many other industrial distributed processes. Modeling is essential to process prediction, control and optimization. Though modeling of lumped parameter systems (LPS) has been widely studied, modeling of DPS, especially with the unknown uncertainties, achieves little progress. This motivates us to study the modeling of unknown nonlinear DPS in this book.

Since the DPS is described in PDE that is infinite-dimensional, even for the known DPS, the reduction to finite-dimensional ordinary differential equations (ODE) is needed because only finite number of actuators/sensors can be used in practice. The common model reduction methods include finite difference method (FDM) (Mitchell & Griffiths, 1980), finite element method (FEM) (Brenner & Ridgway Scott, 1994), spectral method (Canuto *et al.*, 1988; Boyd, 2000) and Karhunen-Loève (KL) method (Sirovich, 1987; Holmes *et al.*, 1996; Newman, 1996a). FDM and FEM may lead to very high-order models, which are not suitable for real-time control. Because the spectral method and KL method can result in low-order models, they are widely used for model reduction in control. However, the spectral method requires more knowledge of the process. Though, a hybrid spectral/intelligent modeling approach is proposed recently (Deng, Li & Chen, 2005) to explore a wider range of application under uncertainties, still it is limited to the parabolic type of processes. Compared with the spectral method, KL method can lead to lower-order models and is suitable for more complex and wider range of systems. However, it requires much more sensors to obtain the process data.

When there are strong uncertainties in the DPS, modeling with the input-output data seems the only solution. If the structure of the DPS is known and only some of its parameters are unknown, then many parameter estimation methods can be used (e.g., Banks & Kunisch, 1989; Coca & Billings, 2000; Demetriou & Rosen, 1994). Though the nonlinear DPS in unknown structure is common in the industry, its modeling problem is extremely difficult and achieves little progress. With little knowledge of the process, a data-based modeling becomes necessary for DPS in unknown structure. These data-based approaches need little physical or chemical knowledge of the process, and thus are more feasible to use in the practice. Still, these kinds of methods have different limitations as follows.

- Green's function based identification (Gay & Ray, 1995; Zheng, Hoo & Piovoso, 2002) leads to a single spatio-temporal kernel model, which is suitable for control design because of its linear structure. However, it is only a linear approximation of the nonlinear DPS. It may not be applicable to a highly nonlinear DPS.
- FDM and FEM based identification (Gonzalez-Garcia, Rico-Martinez & Kevredidis, 1998; Guo & Billings, 2007; Coca & Billings, 2002) will lead to a high-order model, which may result in an impractical high-order controller.
- KL based identification (Zhou, Liu, Dai & Yuan, 1996; Smaoui & Al-Enezi, 2004; Sahan *et al.*, 1997; Romijn *et al.*, 2008; Aggelogiannaki & Sarimveis, 2008; Qi & Li, 2008a) may lead to a low-order model. Since most of KL based identification use neural network for modeling the dynamics, its nonlinear structure may result in a very complicated control design.

In general, it is still very necessary to develop some new unknown nonlinear DPS modeling approach to overcome these limitations.

In modeling of the traditional lumped parameter systems (LPS), the following models have been widely studied because of their significant properties.

- Traditional block-oriented nonlinear models have been often used because of their simple nonlinear structure, ability to approximate a large class of nonlinear processes and efficient control schemes (e.g., Narendra & Gallman, 1966; Stoica & Söderström, 1982; Bai, 1998; Zhu, 2000; Gómez & Baeyens, 2004; Westwick & Verhaegen, 1996; Hagenblad & Ljung, 2000). They consist of the interconnection of linear time invariant (LTI) systems and static nonlinearities. Two common model structures are: the Hammerstein model, which consists of the cascade connection of a static nonlinearity followed by a LTI system, the Wiener model, in which the order of the linear and the nonlinear blocks is reversed.
- Traditional Volterra model consists of a series of temporal kernel, which is a high-order and nonlinear extension of linear impulse response model (Boyd & Chua, 1985; Schetzen, 1980; Rugh, 1981; Doyle III *et al.*, 1995; Maner *et al.*, 1996; Parker *et al.*, 2001). It has been used for LPS modeling because of it simple nonlinear structure.
- In the field of machine learning, nonlinear principal component analysis (NL-PCA) has been widely studied for nonlinear dimension reduction of high-dimensional data (e.g., Dong & McAvoy, 1996; Kramer, 1991; Hsieh, 2001; Kirby & Miranda, 1994; Smaoui, 2004; Webb, 1996; Wilson, Irwin & Lightbody, 1999). They can achieve a lower-order or more accurate model than the KL method for a nonlinear problem because the KL method is only a linear dimension reduction.

However, all these methods are temporal models and can not model spatio-temporal dynamics directly. Whether these methods that are efficient in the traditional temporal dynamics can be applied to the spatio-temporal dynamics is an open question that needs to be answered in this book.

The book has two major objectives:

(1) to develop some useful spatio-temporal modeling approaches for unknown nonlinear DPS, based on the idea of the traditional Wiener, Hammerstein, Volterra and neural method;

(2) to apply the presented methods to thermal processes in IC packaging and chemical industry.

The work in this book mainly includes:

- To provide a systematic overview and classification on the DPS modeling;
- To develop spatio-temporal Wiener and Hammerstein models for the nonlinear DPS with the help of the KL method;
- To develop spatio-temporal Hammerstein and Volterra models for the nonlinear DPS using the idea of spatio-temporal kernels;
- To develop the spatio-temporal neural model for the nonlinear DPS;
- To apply and evaluate the presented models on modeling of typical thermal processes in IC packaging and chemical industry through simulation and experiment.

1.2 Contributions and Organization of the Book

In this book, we will present the spatio-temporal Wiener (Chapter 3), spatio-temporal Hammerstein (Chapters 4 & 5), spatio-temporal Volterra (Chapter 6) and spatio-temporal neural NARX (nonlinear autoregressive with exogenous input) (Chapter 7) modeling approaches for the unknown nonlinear DPS.

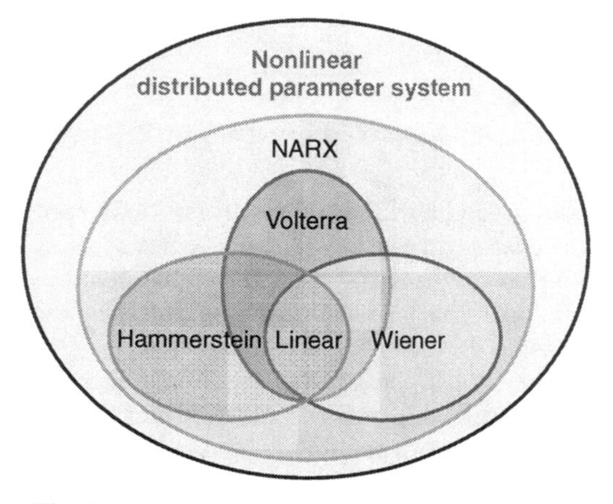

Fig. 1.4 Spatio-temporal models for nonlinear DPS

The relationship of these models is shown in Figure 1.4. The Wiener, Hammerstein and Volterra models are complexity restricted nonlinear models included in a general class of NARX models with the linear model as a special case. The Wiener, Hammerstein and Volterra models have special nonlinear structures, while the NARX model is useful for describing more complex nonlinear systems.

The way to process the spatial information is a crucial work in DPS modeling. In this book, three different approaches are classified in Figure 1.5.

- *KL based approach.* The DPS is separated into a set of spatial basis functions and temporal dynamics using the KL method for time/space separation. Then the traditional modeling approach is used to model the temporal dynamics. With the time-space synthesis, the spatio-temporal dynamics can be reconstructed. The work will be described in details in Chapters 3 and 4.
- *Kernel based approach.* The traditional model can be extended to DPS via the spatio-temporal kernel. After the time/space separation using the KL method, the model can be estimated in the temporal domain. The work will be described in details in Chapters 5 and 6.
- *NL-PCA based approach.* The nonlinear time/space separation of DPS is performed using nonlinear PCA. Then traditional modeling approach can be used to model the temporal dynamics. With the nonlinear time/space synthesis, the spatio-temporal dynamics can be recovered. The work will be described in details in Chapter 7.

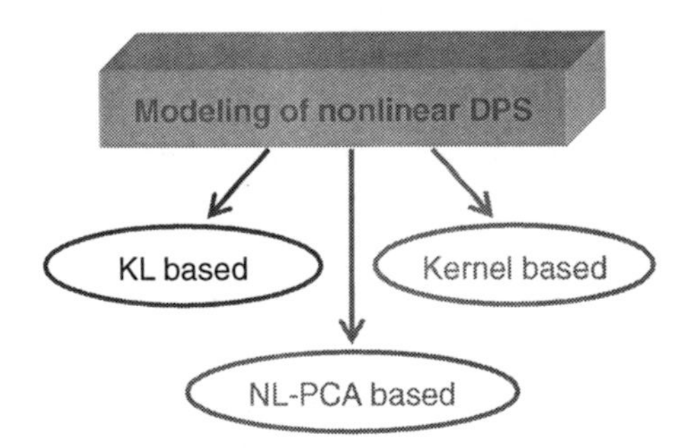

Fig. 1.5 Spatial information processing for DPS modeling

This book provides several useful modeling frameworks for control of unknown nonlinear DPS. The developed modeling methodologies not only can be used in the thermal process, but also can be easily applied to other distributed industrial processes. The contents of each chapter will be summarized as below with emphases on the main contributions.

Chapter 2 is a systematic overview and classification on the DPS modeling. Three DPS modeling problems, model reduction for known DPS, parameter estimation for DPS with unknown parameters, and system identification for DPS in unknown structure, are discussed. Various approaches are classified with their limitations and advantages summarized as well. This overview motivates us to develop new DPS modeling methods.

In Chapter 3, a KL based Wiener modeling approach is presented (Qi & Li, 2008b). A spatio-temporal Wiener model (a linear DPS followed by a static nonlinearity) is established for modeling nonlinear DPS. After the time/space separation, it can be represented by traditional Wiener system with a set of spatial basis functions. To obtain a low-order model, the KL method is used for the time/space separation and dimension reduction. Then the Wiener model is obtained using the least-squares estimation and the instrumental variables method, which can achieve consistent estimates under process noise.

In Chapter 4, a KL based Hammerstein modeling approach is provided (Qi & Li, 2009a). A spatio-temporal Hammerstein model (a static nonlinearity followed by a linear DPS) is constructed for modeling the nonlinear DPS. After the time/space separation, it can be represented by the traditional Hammerstein system with a set of spatial basis functions. To achieve a low-order model, the KL method is used for the time/space separation and dimension reduction. To obtain a parsimonious Hammerstein model, the orthogonal forward regression algorithm is used to determine the compact or sparse model structure, and then the parameters are estimated using the least-squares method and the singular value decomposition. This method can obtain a low-order and parsimonious Hammerstein model.

In Chapter 5, a kernel based multi-channel spatio-temporal Hammerstein modeling approach is presented (Qi, Zhang & Li, 2009). For modeling nonlinear DPS, a spatio-temporal Hammerstein model (a static nonlinearity followed by a linear spatio-temporal kernel) is constructed. A basic identification approach with the least-squares estimation and singular value decomposition can work well if the model structure is matched with the system. When there are some unmodeled dynamics, a multi-channel modeling framework is presented to achieve a better performance, which can guarantee the convergence under noisy measurements.

In Chapter 6, a Volterra kernel based spatio-temporal modeling approach is presented (Li, Qi & Yu, 2009; Li & Qi, 2009). To reconstruct the spatio-temporal dynamics, a spatio-temporal Volterra model is constructed with a series of spatio-temporal kernel. To achieve a low-order model, the KL method is used for the time/space separation and dimension reduction. Then the model is estimated with a least-squares algorithm, which can guarantee the convergence under noisy measurements.

In Chapter 7, a NL-PCA based neural modeling approach is presented (Qi & Li, 2009b). One NL-PCA network is trained for nonlinear dimension reduction and nonlinear time/space reconstruction, and then the other neural model is to learn the system dynamics. With the powerful capability of dimension reduction and intelligent learning, this spatio-temporal neural model can describe more complex and strong nonlinear DPS.

Chapter 8 provides conclusions and future challenges.

References

[1] Aggelogiannaki, E., Sarimveis, H.: Nonlinear model predictive control for distributed parameter systems using data driven artificial neural network models. Computers and Chemical Engineering 32(6), 1225–1237 (2008)
[2] Bai, E.W.: An optimal two-stage identification algorithm for Hammerstein-Wiener nonlinear systems. Automatica 34(3), 333–338 (1998)
[3] Banks, H.T., Kunisch, K.: Estimation techniques for distributed parameter systems. Birkhauser, Boston (1989)
[4] Boyd, J.P.: Chebyshev and Fourier spectral methods, 2nd edn. Dover Publications, New York (2000)
[5] Boyd, S., Chua, L.O.: Fading memory and the problem of approximating nonlinear operators with Volterra series. IEEE Transactions on Circuits and Systems 32(11), 1150–1161 (1985)
[6] Brenner, S.C., Ridgway Scott, L.: The mathematical theory of finite element methods. Springer, New York (1994)
[7] Canuto, C., et al.: Spectral methods in fluid dynamics. Springer, New York (1988)
[8] Christofides, P.D.: Robust control of parabolic PDE systems. Chemical Engineering Science 53(16), 2949–2965 (1998)
[9] Christofides, P.D.: Control of nonlinear distributed process systems: Recent developments and challenges. AIChE Journal 47(3), 514–518 (2001a)
[10] Christofides, P.D.: Nonlinear and robust control of PDE systems: Methods and applications to transport-reaction processes. Birkhäuser, Boston (2001b)
[11] Christofides, P.D., Armaou, A. (eds.): Control of multiscale and distributed process systems - Preface. Computers and Chemical Engineering 29(4), 687–688 (2005)
[12] Christofides, P.D., Wang, X.Z. (eds.): Special issue on 'control of particulate processes'. Chemical Engineering Science 63(5), 1155 (2008)
[13] Christofides, P.D. (ed.): Special volume on 'control of distributed parameter systems'. Computers and Chemical Engineering 26(7-8), 939–940 (2002)
[14] Christofides, P.D. (ed.): Special issue on 'control of complex process systems'. International Journal of Robust and Nonlinear Control 14(2), 87–88 (2004)
[15] Coca, D., Billings, S.A.: Direct parameter identification of distributed parameter systems. International Journal of Systems Science 31(1), 11–17 (2000)
[16] Coca, D., Billings, S.A.: Identification of finite dimensional models of infinite dimensional dynamical systems. Automatica 38(11), 1851–1865 (2002)
[17] Demetriou, M.A., Rosen, I.G.: Adaptive identification of second-order distributed parameter systems. Inverse Problems 10(2), 261–294 (1994)
[18] Deng, H., Li, H.-X., Chen, G.: Spectral-approximation-based intelligent modeling for distributed thermal processes. IEEE Transactions on Control Systems Technology 13(5), 686–700 (2005)
[19] Dochain, D., Dumont, G., Gorinevsky, D., Ogunnaike, T. (eds.): Special issue on 'control of industrial spatially distributed processes'. IEEE Transactions on Control Systems Technology 11(5), 609–611 (2003)
[20] Dong, D., McAvoy, T.J.: Nonlinear principal component analysis-based on principal curves and neural networks. Computers and Chemical Engineering 20(1), 65–78 (1996)
[21] Doyle III, F.J., Ogunnaike, B.A., Pearson, R.K.: Nonlinear model-based control using second-order Volterra Models. Automatica 31(5), 697–714 (1995)

[22] Gay, D.H., Ray, W.H.: Identification and control of distributed parameter systems by means of the singular value decomposition. Chemical Engineering Science 50(10), 1519–1539 (1995)

[23] Gómez, J.C., Jutan, A., Baeyens, E.: Wiener model identification and predictive control of a pH neutralisation process. IEE Proceedings-Control Theory and Applications 151(3), 329–338 (2004)

[24] Gonzalez-Garcia, R., Rico-Martinez, R., Kevredidis, I.G.: Identification of distributed parameter systems: A neural net based approach. Computers and Chemical Engineering 22, S965–S968 (1998)

[25] Guo, L.Z., Billings, S.A.: Sate-space reconstruction and spatio-temporal prediction of lattice dynamical systems. IEEE Transactions on Automatic Control 52(4), 622–632 (2007)

[26] Hagenblad, A., Ljung, L.: Maximum likelihood estimation of Wiener models. Report no.: LiTH-ISY-R-2308, Linköping University, Sweden (2000)

[27] Holmes, P., Lumley, J.L., Berkooz, G.: Turbulence, coherent structures, dynamical systems, and symmetry. Cambridge University Press, New York (1996)

[28] Hsieh, W.W.: Nonlinear principal component analysis by neural networks. Tellus Series A - Dynamic Meteorology and Oceanography 53(5), 599–615 (2001)

[29] Kirby, M., Miranda, R.: The nonlinear reduction of high-dimensional dynamical systems via neural networks. Physical Review Letter 72(12), 1822–1825 (1994)

[30] Kramer, M.A.: Nonlinear principal component analysis using autoassociative neural networks. AIChE Journal 37(2), 233–243 (1991)

[31] Li, H.-X., Qi, C.K.: Incremental modeling of nonlinear distributed parameter processes via spatio-temporal kernel series expansion. Industrial & Engineering Chemistry Research 48(6), 3052–3058 (2009)

[32] Li, H.-X., Qi, C.K., Yu, Y.G.: A spatio-temporal Volterra modeling approach for a class of nonlinear distributed parameter processes. Journal of Process Control 19(7), 1126–1142 (2009)

[33] Maner, B.R., Doyle III, F.J., Ogunnaike, B.A., Pearson, R.K.: Nonlinear model predictive control of a simulated multivariable polymerization reactor using second-order Volterra models. Automatica 32(9), 1285–1301 (1996)

[34] Mitchell, A.R., Griffiths, D.F.: The finite difference method in partial differential equations. Wiley, Chichester (1980)

[35] Narendra, K., Gallman, P.: An iterative method for the identification of nonlinear systems using a Hammerstein model. IEEE Transactions on Automatic Control 11(3), 546–550 (1966)

[36] Newman, A.J.: Model reduction via the Karhunen-Loève expansion part I: An exposition. Technical Report T.R.96-32, University of Maryland, College Park, Maryland (1996a)

[37] Parker, R.S., Heemstra, D., Doyle III, F.J., Pearson, R.K., Ogunnaike, B.A.: The identification of nonlinear models for process control using tailored "plant-friendly" input sequences. Journal of Process Control 11(2), 237–250 (2001)

[38] Qi, C.K., Li, H.-X.: Hybrid Karhunen-Loève/neural modeling for a class of distributed parameter systems. International Journal of Intelligent Systems Technologies and Applications 4(1-2), 141–160 (2008a)

[39] Qi, C.K., Li, H.-X.: A Karhunen-Loève decomposition based Wiener modeling approach for nonlinear distributed parameter processes. Industrial & Engineering Chemistry Research 47(12), 4184–4192 (2008b)

[40] Qi, C.K., Li, H.-X.: A time/space separation based Hammerstein modeling approach for nonlinear distributed parameter. Computers & Chemical Engineering 33(7), 1247–1260 (2009a)

[41] Qi, C.K., Li, H.-X.: Nonlinear dimension reduction based neural modeling for spatio-temporal processes. Chemical Engineering Science 64(19), 4164–4170 (2009b)

[42] Qi, C.K., Zhang, H.-T., Li, H.-X.: A multi-channel spatio-temporal Hammerstein modeling approach for nonlinear distributed parameter processes. Journal of Process Control 19(1), 85–99 (2009)

[43] Romijn, R., Özkan, L., Weiland, S., Ludlage, J., Marquardt, W.: A grey-box modeling approach for the reduction of nonlinear systems. Journal of Process Control 18(9), 906–914 (2008)

[44] Rugh, W.: Nonlinear system theory: The Volterral/Wiener approach. Johns Hopkins University Press, Baltimore (1981)

[45] Sahan, R.A., Koc-Sahan, N., Albin, D.C., Liakopoulos, A.: Artificial neural network-based modeling and intelligent control of transitional flows. In: Proceeding of the 1997 IEEE International Conference on Control Applications, Hartford, CT, pp. 359–364 (1997)

[46] Schetzen, M.: The Volterra and Wiener theories of nonlinear systems. Wiley, New York (1980)

[47] Sirovich, L.: Turbulence and the dynamics of coherent structures parts I-III. Quarterly of Applied Mathematics 45(3), 561–590 (1987)

[48] Smaoui, N.: Linear versus nonlinear dimensionality reduction of high-dimensional dynamical systems. SIAM Journal on Scientific Computing 25(6), 2107–2125 (2004)

[49] Smaoui, N., Al-Enezi, S.: Modelling the dynamics of nonlinear partial differential equations using neural networks. Journal of Computational and Applied Mathematics 170(1), 27–58 (2004)

[50] Stoica, P., Söderström, T.: Instrumental-variable methods for identification of Hammerstein systems. International Journal of Control 35(3), 459–476 (1982)

[51] Webb, A.R.: An approach to non-linear principal components analysis using radially symmetric kernel functions. Journal Statistics and Computing 6(2), 159–168 (1996)

[52] Westwick, D., Verhaegen, M.: Identifying MIMO Wiener systems using subspace model identification methods. Signal Processing 52(2), 235–258 (1996)

[53] Wilson, D.J.H., Irwin, G.W., Lightbody, G.: RBF principal manifolds for process monitoring. IEEE Transactions on Neural Networks 10(6), 1424–1434 (1999)

[54] Zheng, D., Hoo, K.A., Piovoso, M.J.: Low-order model identification of distributed parameter systems by a combination of singular value decomposition and the Karhunen-Loève expansion. Industrial & Engineering Chemistry Research 41(6), 1545–1556 (2002)

[55] Zhou, X.G., Liu, L.H., Dai, Y.C., Yuan, W.K., Hudson, J.L.: Modeling of a fixed-bed reactor using the KL expansion and neural networks. Chemical Engineering Science 51(10), 2179–2188 (1996)

[56] Zhu, Y.C.: Identification of Hammerstein models for control using ASYM. International Journal of Control 73(18), 1692–1702 (2000)

2 Modeling of Distributed Parameter Systems: Overview and Classification

Abstract. This chapter provides a systematic overview of the distributed parameter system (DPS) modeling and its classification. Three different problems in DPS modeling are discussed, which includes model reduction for known DPS, parameter estimation for DPS, and system identification for unknown DPS. All approaches are classified into different categories with their limitations and advantages briefly discussed. This overview motivates us to develop new methods for DPS modeling.

2.1 Introduction

When the distributed parameter system (DPS) is known, partial differential equations (PDE), which are derived from the physical and chemical knowledge (i.e., *first-principle modeling*) under simplified assumptions, can provide a nominal model of the system. Though there are usually some uncertainties between the nominal model and the system, it can capture the dominant dynamics of the system. Because of limited computation capacity for numerical implementation and a finite number of actuators and sensors for sensing and control, such infinite-dimensional systems need to be reduced into finite-dimensional approximation systems, e.g., ordinary differential equations (ODE), and difference equations (DE), etc. This is so called *model reduction* of the DPS.

However, in many situations it is very difficult to get an accurate nominal PDE description via the first-principle modeling because of incomplete physical and chemical process knowledge (e.g., unknown system parameters, unknown model structure and disturbance). These uncertainties existing in the process make the modeling problem more difficult and challenging.

When the DPS is unknown, extra efforts are needed besides the previous model reduction. Two different situations may happen. One is *grey-box modeling*, where the PDE structure is available from a priori knowledge with only some parameters need to determine. These unknown parameters can be estimated from the input-output data, which is the so-called *parameter estimation* of the DPS. Another situation is *black-box modeling*, where the PDE structure is unknown. Then both model structure and parameters need to be determined or identified from the measurement, which is the so-called *system identification* of the DPS.

The aforementioned problems: *model reduction*, *parameter estimation* and *system identification* are fundamentals of the DPS modeling. In the last several decades, many researchers in the field of mathematics and engineering have made much effort on these problems. For each problem, though many different methods have been developed, most of these methods can be synthesized into several

H.-X. Li and C. Qi: Spatio-Temporal Modeling of Nonlinear DPS, ISCA 50, pp. 13–49.
springerlink.com

categories. Moreover, different problems and their corresponding methods perhaps share some common properties. Though this book will focus on the system identification problem, for easy understanding, it will only provide an overview on selected problems rather than the whole area.

Since it is a very large and complex field, it is almost impossible to review every work in the area. We will focus on the applicable methods and their classifications. For simplicity, some of the complicated mathematical theories, such as, existence and uniqueness of the PDE solution, modeling error analysis, parameter identifiability, convergence of the estimation algorithm, etc., will not be included in this chapter. See the related references for more details on these problems.

This chapter is organized as follows: model reduction for known DPS in Section 2.2, parameter estimation for known structure of DPS in Section 2.3 and system identification for unknown structure of DPS in Section 2.4. Conclusions will be given in Section 2.5 together with new modeling ideas that will be studied in the rest of the book.

For simplicity, we consider a class of quasi-linear parabolic systems in one-dimensional spatial domain

$$\frac{\partial y(x,t)}{\partial t} = \alpha \frac{\partial^2 y}{\partial x^2} + \beta \frac{\partial y}{\partial x} + f(y) + wb(x)u(t) , \tag{2.1}$$

subject to the boundary conditions

$$y(0,t) = 0, y(\pi,t) = 0 , \tag{2.2}$$

and the initial condition

$$y(x,0) = y_0(x) , \tag{2.3}$$

where $y(x,t) \in \mathbb{R}$ denotes the output variable, $[0,\pi] \subset \mathbb{R}$ is the spatial domain of definition of the system, $x \in [0,\pi]$ is the spatial coordinate, $t \in [0,\infty)$ is the time, $u(t) = [u_1 \ u_2 \ \dots \ u_m]^T \in \mathbb{R}^m$ denotes the vector of manipulated inputs, $f(y)$ is a nonlinear function, α , β and w are constants, $b(x)$ is a vector function of x of the form $b(x) = [b_1 \ b_2 \ \dots \ b_m]$, $b_i(x)$ describes how the control action $u_i(t)$ is distributed in the interval $[0,\pi]$. Define the spatial operator $\mathcal{A}$ as

$$\mathcal{A}y = \alpha \frac{\partial^2 y}{\partial x^2} + \beta \frac{\partial y}{\partial x} . \tag{2.4}$$

Two industrial cases will be used to demonstrate and compare different modeling methods (e.g., spectral method and KL method). The first case is a quasi-linear parabolic process, and the second one is a nonlinear parabolic process.

Case 1: Quasi-linear Parabolic Process

Consider the catalytic rod given in Section 1.1.2, and assume that the dimensionless reaction rate β_T is spatially-varying, the spatio-temporal evolution of the dimensionless rod temperature is described by the following quasi-linear parabolic PDE,

$$\frac{\partial y}{\partial t} = k\frac{\partial^2 y}{\partial x^2} + \beta_T(x)(e^{-\frac{\gamma}{1+y}} - e^{-\gamma}) + \beta_u(b(x)u(t) - y), \quad (2.5)$$

subject to the Dirichlet boundary conditions,

$$y(0,t) = 0, y(\pi,t) = 0, \quad (2.6)$$

and the initial condition,

$$y(x,0) = y_0(x), \quad (2.7)$$

where the parameter values are given in Table 2.1.

Table 2.1 Dimensionless parameters for Case 1

Parameter	Real values	Nominal values	Definition
k	1	1	Diffusion coefficient
β_T	$12(\cos(x)+1)$	---	Heat of reaction
β_u	20	20	Heat transfer coefficient
γ	2	---	Activation energy

Case 2: Nonlinear Parabolic Process

Consider the catalytic rod given in Section 1.1.2 again, and assume that the spatial differential operator is nonlinear (e.g., nonlinear dependence of the thermal conductivity on temperature) with the convection feature and that the dimensionless reaction rate β_T is spatially-varying. In this case, the process will be described in the following nonlinear parabolic PDE,

$$\frac{\partial y}{\partial t} = \frac{\partial}{\partial x}(k(y)\frac{\partial y}{\partial x}) - v\frac{\partial y}{\partial x} + \beta_T(x)(e^{-\frac{\gamma}{1+y}} - e^{-\gamma}) + \beta_u(b(x)u(t) - y), \quad (2.8)$$

subject to the boundary conditions,

$$y(0,t) = 0, y(\pi,t) = 0, \quad (2.9)$$

and the initial condition,

$$y(x,0) = y_0(x), \quad (2.10)$$

where the parameter values can be found in Table 2.2.

Table 2.2 Dimensionless parameters for Case 2

Parameter	Real values	Nominal values	Definition
k	$1+0.7/(y+1)$	1	Diffusion coefficient
v	10	10	
β_T	$12(\cos(x)+1)$	---	Heat of reaction
β_u	20	20	Heat transfer coefficient
γ	2	---	Activation energy

2.2 White-Box Modeling: Model Reduction for Known DPS

Rigorous mathematical models could be derived from the first-principle knowledge under simplified conditions, but they may involve unknown parameters. We will review the parameter estimation problem in Section 2.3. Here, we assume that the PDE description of the DPS is completely known. Then the simulation and control of the DPS are usually accomplished by transforming the PDE and boundary conditions into a finite-dimensional system, such as ordinary differential equation or difference equation (DE). These lumping (i.e., model reduction) methods include: eigenfunction method, Green's function method, finite difference method, and a class of weighted residual method (e.g., Galerkin, collocation, finite element, spectral and Karhunen-Loève method).

2.2.1 Eigenfunction Method

The analytical solution of linear partial differential equations could be obtained using the method of separation of variables (Powers, 1999). First, assume the solution can be expressed by the following form of separation of variables

$$y(x,t)=\sum_{i=1}^{\infty}\phi_i(x)y_i(t)=\overline{\phi}^T(x)\overline{y}(t), \tag{2.11}$$

where $\overline{\phi}(x)$ and $\overline{y}(t)$ are the corresponding spatial and temporal vectors, respectively. Substituting it into the original linear PDE will yield the eigenvalues and eigenfunctions. For $\mathcal{A}$, the standard eigenvalue and eigenfunction problem is of the form

$$\mathcal{A}\phi_j=\lambda_j\phi_j, j=1,\ldots,\infty, \tag{2.12}$$

where λ_j denotes an eigenvalue and ϕ_j denotes an eigenfunction. For many typical operator $\mathcal{A}$ and boundary conditions, the eigenvalue and eigenfunction can be found from the book (Butkovskiy, 1982). It is an infinite-dimensional eigenfunction solution. By truncation, a finite-dimensional approximation solution will be obtained.

A low-order model may be possible only if the system is self-adjoint. Otherwise, a high-order solution should be used for a satisfactory accuracy. High-order models always result in a difficult control realization. This is primarily attributed to the lack of measurement for characterizing the distributed nature and the lack of actuators for control implementation.

Moreover, the condition of separation of variables is very rigor. Whether or not the method of separation of variables can be applied to a particular problem depends on not only the differential equation but also on the shape of the domain and the form of the boundary conditions.

In particular, for a nonlinear PDE, it is difficult to separate the variables and find the set of analytical eigenfunctions. One approach is to linearize the nonlinear system at a nominal state using the Taylor series expansion. The resulting system is a linear system plus nonlinear terms. The eigenfunctions of linear part can be used to transfer the PDE to an ODE. Rigorously speaking, this method is different from the original method of separation of variables, and in fact it belongs to a new method, i.e., weighted residual method, which will be discussed in Section 2.2.4.

2.2.2 Green's Function Method

The characteristic of linear DPS can be completely represented by the *Green's function* as below (also called the impulse response function or kernel).

$$y(z,t) = \int_0^t \int_0^\pi g(z,\zeta,t-\tau)u(\zeta,\tau)d\zeta d\tau \,. \tag{2.13}$$

If the eigenvalue and eigenfunction can be solved analytically, the Green's function can be expressed by an infinite-dimensional eigenfunction expansion analytically. By truncation, a finite-dimensional approximation can be obtained. The Green's functions of many typical linear differential operators have been given in (Butkovskiy, 1982). This approach assumes the PDE is linear, though most of DPS are essentially nonlinear. Furthermore, a low-order eigenfunction representation of Green's function may be possible only if the system is self-adjoint. However, there are no assurances that they are self-adjoint. Fortunately, for the non-self-adjoint system, the singular function representation of Green's function could be low-order, and can be estimated from the input-output data. More details are introduced in Section 2.4.1.

2.2.3 Finite Difference Method

Finite difference method (FDM) is a popular method to provide the numerical solution of the PDE (Mitchell & Griffiths, 1980). The spatio-temporal variables are discretized within the time-space domain as illustrated in Figure 2.1a. Derivatives at each discretization node are approximated by the difference over a small interval, which can be forward, backward and central difference derived often from Taylor expansion. Then the PDE system is transformed into a set of difference equations, whose order is proportional to the number of spatial discretization nodes. The most attractive feature of FDM is that it can work for all kinds of DPS with various boundary conditions and regular domain. However, it usually requires a high-order model for an accurate solution, and has the disadvantages of heavy computational burden.

The *method of lines* (MOL) (Schiesser, 1991) is a special case of the FDM, with only partial derivatives in the spatial direction replaced by finite difference approximations, as illustrated in Figure 2.1b. This results in a system of ODE because of the spatial discretization. It has the similar strength and weakness as FDM.

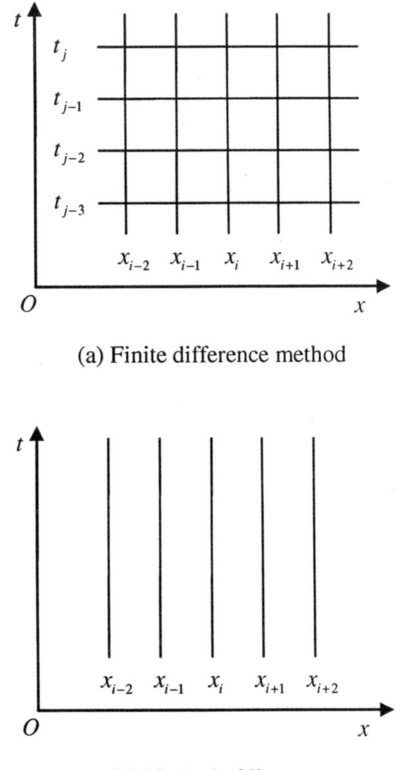

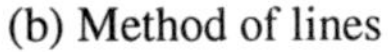
(b) Method of lines

Fig. 2.1 Geometric interpretations of finite difference and method of lines

2.2.4 Weighted Residual Method

The *weighted residual method* (WRM) (Ray, 1981; Fletcher, 1984) is the most often used and most efficient lumping method for DPS. It is well known that a continuous function can be approximated using Fourier series (Zill & Cullen, 2001). Based on this principle, the spatio-temporal variable $y(x,t)$ of the DPS can be expanded by a set of spatial basis functions (BFs) $\{\phi_i(x)\}_{i=1}^{\infty}$ as follows

$$y(x,t)=\sum_{i=1}^{\infty}\phi_i(x)y_i(t)=\bar{\phi}^T(x)\bar{y}(t)\,, \tag{2.14}$$

where $\bar{\phi}(x)$ and $\bar{y}(t)$ are the corresponding vectors. Similar to Fourier series, the spatial BFs are often ordered from slow to fast in the spatial frequency domain. Because the fast modes contribute little to the whole system, only the first n slow modes in the expansion will be retained in practice (Fletcher, 1984)

$$y_n(x,t)=\sum_{i=1}^{n}\phi_i(x)y_i(t)=\bar{\phi}_n^T(x)\bar{y}_n(t)\,, \tag{2.15}$$

where $\bar{\phi}_n(x)$ and $\bar{y}_n(t)$ are the corresponding vectors. Thus, the spatio-temporal variable is separated into a set of spatial BFs and the temporal variables, as depicted in Figure 2.2. The key is to select proper spatial BFs, and construct the finite-order (low-order) temporal model as explained in Figure 2.3. Finally, through the time-space synthesis, the spatio-temporal system will be recovered.

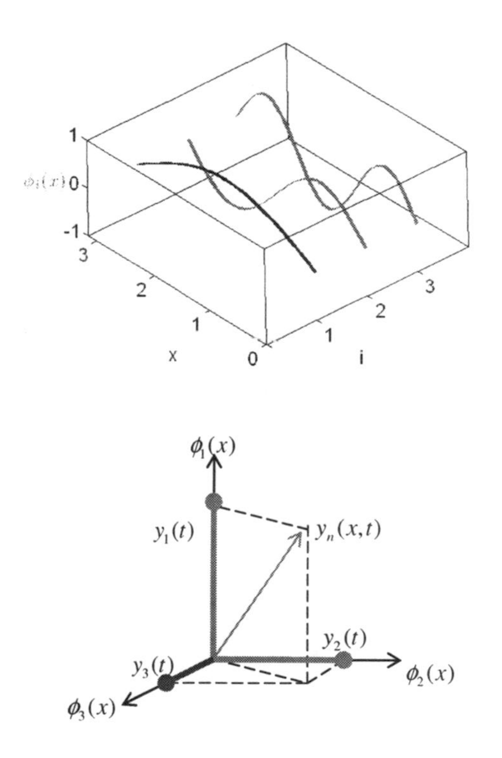

Fig. 2.2 Geometric interpretation of time-space separation for n=3

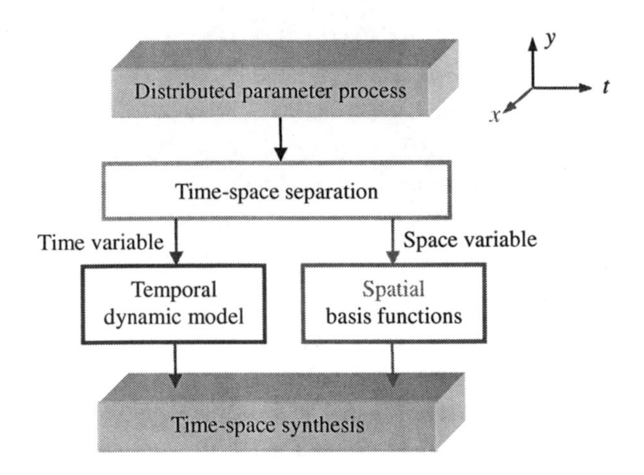

Fig. 2.3 Framework of weighted residual method

In the WRM, the equation residual of the model (2.1) generated from the truncated expansion can be expressed as

$$R(x,t) = \dot{y}_n - (\alpha \frac{\partial^2 y_n}{\partial x^2} + \beta \frac{\partial y_n}{\partial x} + f(y_n) + wbu), \tag{2.16}$$

which is made small in the sense that

$$(R, \varphi_i) = 0, i = 1,...,n, \tag{2.17}$$

where $\{\varphi_i(x)\}_{i=1}^n$ are a set of weighting functions to be chosen. As shown in Figure 2.4, the minimization of the residual R actually turns to minimize its projections onto weighting functions. This is an easy way to obtain a n-order ODE model for $\{y_i(t)\}_{i=1}^n$.

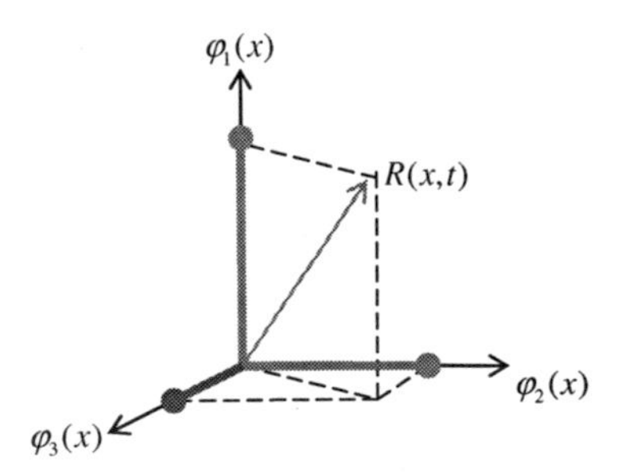

Fig. 2.4 Geometric interpretation of weighted residual method

Notice that WRM is an extension of the eigenfunction method with the difference that WRM may use any of basis functions, while the eigenfunction method uses the eigenfunctions of the linear operator. It is possible to apply WRM to both linear and nonlinear PDE systems.

The accuracy and efficiency of WRM is very dependent on the basis and weighting functions chosen (Fletcher, 1984). Thus, in the following sections, WRM will be further classified into different methods according to the type of weighting and basis functions used. Of course, all these methods are the combination of temporal model construction and spatial BFs selection.

2.2.4.1 Classification Based on Weighting Functions

Many methods have been proposed based on the selection of weighting functions. The most popular approach appears to be *Galerkin method* and *collocation method.*

2.2.4.1.1 Galerkin Method

If the weighting functions $\{\varphi_i(x)\}_{i=1}^n$ are chosen to be the BFs $\{\phi_i(x)\}_{i=1}^n$, then the method is called the *Galerkin method* (Ray, 1981; Fletcher, 1984). It has the advantage that the residual is made orthogonal to each BF and is, therefore, the best solution possible in the space made up of the n functions $\phi_i(x)$. Because it does not need to find other weighting functions, this method is relatively simple and most often used.

Consider the PDE system described by (2.1)-(2.3), define the first finite modes $\overline{y}_n(t) = [y_1(t), y_2(t),..., y_n(t)]^T$ as slow modes and the last infinite modes $\overline{y}_r(t) = [y_{n+1}(t), y_{n+2}(t),..., y_\infty(t)]^T$ as fast modes, using the Galerkin method yields

$$\frac{d\overline{y}_n}{dt} = A_r\overline{y}_n + A_{nr}\overline{y}_r + F_n(\overline{y}_n, \overline{y}_r) + B_n u \,, \tag{2.18}$$

$$\frac{d\overline{y}_r}{dt} = A_{rn}\overline{y}_n + A_r\overline{y}_r + F_r(\overline{y}_n, \overline{y}_r) + B_r u \,, \tag{2.19}$$

$$y = C_n\overline{y}_n + C_r\overline{y}_r, \ \overline{y}_n(0) = 0, \ \overline{y}_r(0) = 0 \,, \tag{2.20}$$

where A_n, A_r, A_{nr}, A_{rn}, F_n, F_r, B_n, B_r, C_n and C_r are proper dimensional matrix (see Balas, 1983; Christofides, 2001b). Even though the basis functions are orthonormal, the terms A_{nr} and A_{rn} may be nonzero because of spill-over effects (Balas, 1986). The spill-over means that those residual modes have effects on the dominant modes (modes coupling). It is true that $(\phi_i, \phi_j) = \delta_{ij}$, but it is typically not true that $(\phi_i, A\phi_j) = \delta_{ij}$. That means $A_{nr} \neq 0$ and $A_{rn} \neq 0$.

Because the fast modes contribute very little to the whole system, after neglecting the residual modes the following finite-dimensional nonlinear system is obtained

$$\frac{d\overline{y}_n}{dt} = A_r\overline{y}_n + F_n(\overline{y}_n, 0) + B_n u \,, \tag{2.21}$$

$$y = C_n\overline{y}_n,\ \overline{y}_n(0) = 0 \,, \tag{2.22}$$

This approach is referred to the linear Galerkin method because the residual modes are completely ignored.

2.2.4.1.2 Collocation Method

The weighting functions of the *collocation method* (Ray, 1981) are chosen to be Dirac delta functions $\delta(x - x_i)$, $i = 1,...,n$. The residuals vanish at collocation nodes $\{x_i\}_{i=1}^{n}$, i.e. $R(x_i) = 0$, so the collocation nodes are very critical for the modeling performance. Fortunately, some mathematical theories show that they can be specified automatically in an optimal way. For example, they can be chosen as the roots of the orthogonal polynomials (e.g. Lefèvre *et al.*, 2000; Dochain *et al.*, 1992). If the orthogonal functions are used as the BFs, it is so called the *orthogonal collocation method* (Ray, 1981).

Both Galerkin and collocation methods are linear reduction methods that work fine for the linear DPS. Since the fast modes $\{y_i(t)\}_{i=n+1}^{\infty}$ are completely ignored, some information of the slow modes may also get lost for the nonlinear DPS because of the coupling between the slow and fast modes. To improve the model accuracy and avoid a high-order model at the same time, *nonlinear reduction methods* are often used for the nonlinear DPS modeling, where some fast modes $\{y_i(t)\}_{i=n+1}^{n+m}$ are compensated as a function of the slow modes $\{y_i(t)\}_{i=1}^{n}$. One form of *nonlinear reduction method* is based on *inertial manifold* (IM) (Temam, 1988), where the fast modes are accurately described by the slow modes, and then the DPS can be exactly transformed into a finite-dimensional system. However, for many nonlinear DPS, IM may not exist or may be difficult to find.

Remark 2.1: Approximated Inertial Manifold Method

To overcome the above problems, the *approximated inertial manifold* (AIM) is used to approximately compensate the fast modes with the slow modes (Foias, Jolly *et al.*, 1988; Shvartsman & Kevrikidis, 1998; Christofides & Daoutidis, 1997). Even for the systems with unknown existence of IM, the AIM method can often achieve a better performance than the linear Galerkin or collocation methods.

A general expression of fast modes $\hat{\overline{y}}_r$ is given by

$$\hat{\overline{y}}_r = G(\overline{y}_n(t), u) \,, \tag{2.23}$$

With substitution of (2.23) into (2.18), a finite-dimensional model is given by

$$\frac{d\overline{y}_n}{dt} = A_r\overline{y}_n + A_{nr}\hat{\overline{y}}_r + F_n(\overline{y}_n, \hat{\overline{y}}_r) + B_n u \,. \tag{2.24}$$

An application of the AIM method can increase the accuracy of the reduced-order system without increasing the order of the basis functions.

The key problem is to obtain an approximation of G. There are several approaches to obtain the AIM.

- To assume that the fast modes are at pseudo-steady state, then the so-called *steady manifold* (Foias & Temam, 1988) can approximate IM easily, by ignoring the dynamic information of the fast modes.
- To consider the dynamic information, another AIM is used by integrating the fast modes for a short time using an *implicit Euler method* (Foias, Jolly *et al.*, 1988).
- To further improve the approximation accuracy, a novel procedure based on *singular perturbations method* can be used to construct the AIM with an arbitrary accuracy under certain conditions (Christofides & Daoutidis, 1997; Christofides, 2001b).

All the above AIM can be implemented in either the Galerkin or the collocation approach, which leads to *Galerkin AIM* and *collocation AIM* methods. There are other means of obtaining the AIM, such as, finite difference, which is not discussed here.

Remark 2.2: ***Other Weighting Functions***

Note that the weighting functions are not limited to previous choices. For example, in method of subdomains, the weighting functions are chosen to be a set of Heaviside functions breaking the region into subdomain; in method of moments, the weighting functions are chosen to be powers of x.

2.2.4.2 Classification Based on Basis Functions

Selection of spatial BFs is critical to the WRM. It will have a great impact to the modeling performance. As shown in Table 2.3, the spatial BFs can be classified into *local*, and *global* types, and further into *analytical* and *data-based* functions. In general, there are three major approaches, the *finite element method* (FEM), the *spectral* method and the *Karhunen-Loève* (KL) method.

Table 2.3 Classification of spatial basis functions

Type of BF	Analytical	Data-based
Local $\phi(x)$ (plot of local basis functions over x)	FEM • Low-order piecewise polynomials • Splines • Wavelets, …, etc.	

Table 2.3 (*continued*)

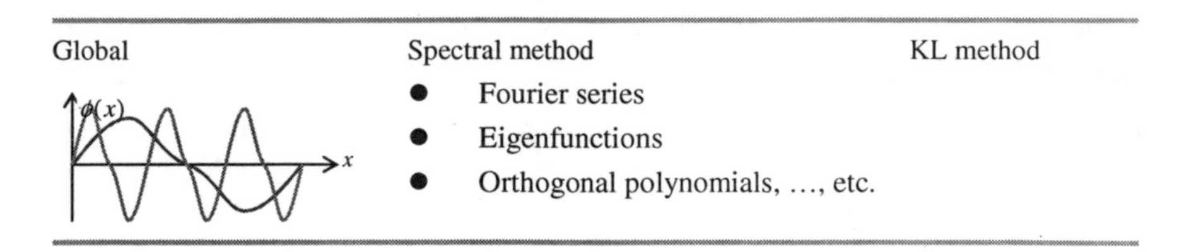

Global	Spectral method	KL method
$\phi(x)$, x	• Fourier series • Eigenfunctions • Orthogonal polynomials, …, etc.	

The data-based *KL* BFs are often global; and no data-based local BFs have been reported so far. Thus, rigorously speaking, KL method can also be considered as a kind of basis functions in spectral method. However, it is convenient to discuss them separately because they have some important differences. Different model reduction approach will be generated when different spatial BFs work with the Galerkin, collocation and AIM method.

2.2.4.2.1 Finite Element Method

The spatial BFs of the FEM are local. The spatial domain is first discretized into sub-domains. Then the *low-order piecewise polynomials* (Brenner & Ridgway Scott, 1994), *splines* (Höllig, 2003), and *wavelets* (Ko, Kurdila & Pilant, 1995; Mahadevan & Hoo, 2000) are often used as local BFs in subdomains. The number of basis functions is determined by the number of discretization points. Figure 2.5 shows an example of piecewise linear polynomials in one dimension.

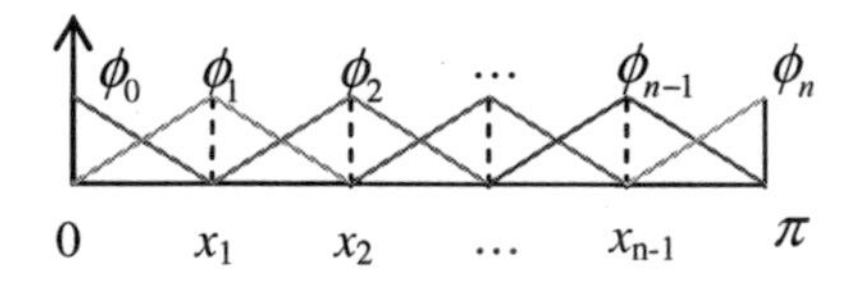

Fig. 2.5 Piecewise linear polynomials in one dimension

Further combined with temporal model construction methods, different applications are reported.

- Examples include, the *spline-Galerkin* method for the order reduction of the controller (Balas, 1986), the *wavelet-Galerkin* method and the *wavelet-Galerkin-AIM* method (Mahadevan & Hoo, 2000) for the reduced-order modeling.
- Other examples include, the *piecewise polynomials-Galerkin* method for an AIM implementation (Marion & Temam, 1990), the *wavelet-collocation* method for numerical simulation (Cruz *et al.*, 2001), and the *wavelet-collocation-AIM* method for the reduced-order modeling (Adrover *et al.*, 2000 & 2002).

The most attractive feature of the FEM is its flexible ability to handle complex geometries and boundaries because of the local BFs used. Due to its flexibility, many FEM software products are developed for the DPS simulation. Because of the local BFs, the FEM often requires a high-order model for a good approximation. The higher the order is, the larger the computational burden will be. Other disadvantages of a high-order solution include stability analysis and synthesis of implemental controllers. For the purpose of control synthesis for the DPS, it is reasonable to seek a methodology that yields a finite-order and accurate enough approximation model.

Both FDM and MOL can fall into the framework of FEM, if their spatial BFs and weighting functions are both chosen as Dirac delta functions. Thus, the FDM and MOL can be viewed as a subset of the FEM approach. The differences between FEM and FDM are:

(1) The FDM is an approximation to the differential equation; the FEM is an approximation to its solution.

(2) The FEM is easy to handle complex geometries (and boundaries) with relative ease, and FDM in its basic form is restricted to handle rectangular shapes and simple alterations.

(3) The most attractive feature of FDM is its ease in implementation.

(4) The quality of a FEM approximation is often higher than that of the corresponding FDM approach, because the quality of the approximation between grid points is poor in FDM.

2.2.4.2.2 Spectral Method

The spatial BFs of the *spectral method* are global and orthogonal in the whole spatial domain (Canuto *et al.*, 1988; Boyd, 2000). Due to the global nature of BFs, the spectral method can achieve a lower order model than the FEM. However, for the same reason, an efficient spectral method often requires that the system has a regular space domain and smooth output. In particular, most of parabolic systems have the spectral gap between the slow and fast modes, thus it is possible to derive an accurate low-order model using the spectral method. To obtain a satisfactory model, the BFs should be carefully designed according to some practical situations such as boundary condition and space domain. Some typical BFs are discussed as follows.

Fourier Series

Fourier series has some good properties such as approximation capability, infinite differential and periodic functions. So it is often used for the processes with periodic boundary conditions and a finite domain (Boyd, 2000; Canuto *et al.*, 1988).

Eigenfunctions

The eigenfunctions (EF) of the linear or linearized spatial operator can be selected as basis functions. For a spatial operator $\mathcal{A}$ in (2.4), the eigenvalue and eigenfunction problem is of the form

$$\mathcal{A}\phi_i = \lambda_i\phi_i, i = 1,\ldots,\infty \,, \tag{2.25}$$

where λ_i denotes the ith eigenvalue and ϕ_i denotes the ith eigenfunction. Figure 2.6 shows the first four eigenfunctions of Case 1: $\phi_i = \sqrt{2/\pi}\sin(iz)$, ($i = 1,\ldots,4$). Eigenfunctions are suitable for most of parabolic PDE systems (Christofides, 1998; Christofides & Baker, 1999) because their eigenspectrum display a separation of the eigenvalues, i.e. slow and fast modes as shown in Figure 2.7. In this situation, an accurate model with a finite number of eigenfunctions is possible.

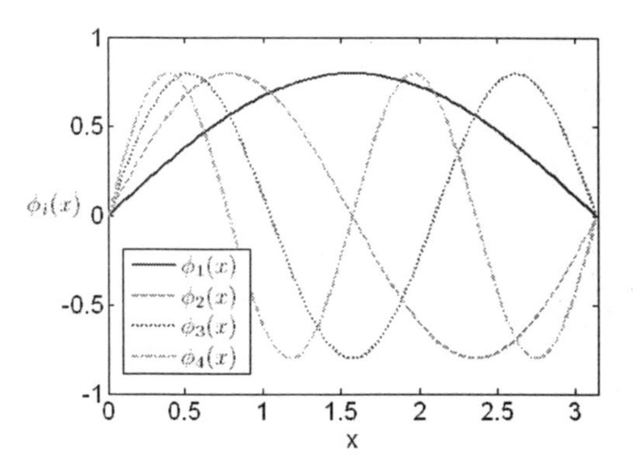

Fig. 2.6 Eigenfunctions of Case 1

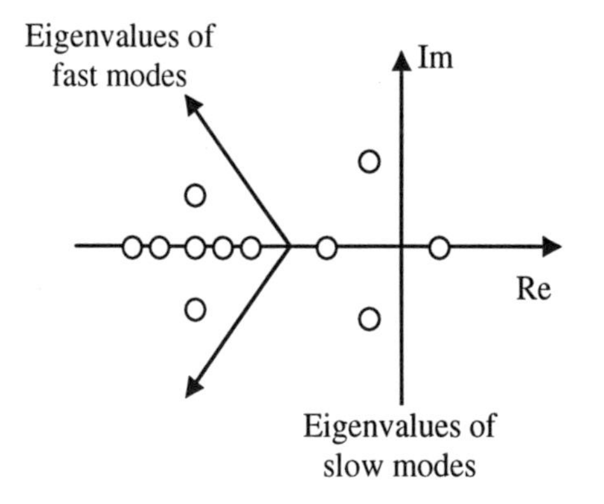

Fig. 2.7 Separation of eigenspectrum

Many control applications are based on the finite-order ODE models derived from the *EF-Galerkin* method, especially for quasi-linear parabolic systems. For example, control of a class of quasi-linear parabolic diffusion-reaction processes (Dubljevic, Christofides & Kevrekidis, 2004), and control of more complex distributed systems, which include the cases of parameter uncertainties (Armaou & Christofides, 2001b; Christofides & Baker, 1999) and input constraints (El-Farra, Armaou & Christofides, 2003; El-Farra & Christofides, 2004). To further improve the modeling performance, the *EF-Galerkin-AIM* method is reported to control quasi-linear parabolic processes (Christofides & Daoutidis, 1997), the process in time-dependent spatial domains (Armaou & Christofides, 1999), and the process with parameter uncertainties (Christofides, 1998).

Except the above successful applications, the EF method also has its limitations.

- When the spatial operator is self-adjoint (Gay & Ray, 1995), the model may be low-order. However, for the non-self-adjoint system, the resulting model is of higher order or even unstable because of the slow convergence of the EF solution (Gay & Ray, 1995; Mahadevan & Hoo, 2000; Hoo & Zheng, 2001).
- For many typical operators and regular boundary conditions, the eigenvalue and EF can be easily derived (Butkovskiy, 1982). However, for the system with nonlinear spatial operators, spatially varying parameters, complex boundary conditions and irregular domain, it will be very difficult and even impossible to get analytical EF.
- EF may be hardly applied to hyperbolic systems, because they do not show the clear separated eigenspectrum. Thus, the *method of characteristics* (Ray, 1981) might be a good choice, where the system is transformed into a set of ODEs, which describe the original DPS along with their characteristic lines.

Orthogonal Polynomials

Orthogonal polynomials are also popular in model reduction of the DPS. For example, the *polynomial-collocation* method is used to derive a low-order DPS model for the purpose of simulation and control (Lefèvre *et al.*, 2000), and for the adaptive control (Dochain *et al.*, 1992). In general, *Chebyshev polynomials* (Boyd, 2000; Canuto *et al.*, 1988) and *Legendre polynomials* (Canuto *et al.*, 1988) are suitable for non-periodic problem defined on a finite domain. *Laguerre polynomials* work well on the semi-infinite domain. *Hermite polynomials* (Boyd, 2000) are useful for the problems with an infinite domain. These functions are very flexible and can be applicable to a broad class of systems. However, they may not be optimal since these general polynomials do not utilize any specific knowledge about the system.

2.2.4.2.3 Karhunen-Loève Method

The KL method, also known as proper orthogonal decomposition (POD) or principal component analysis (PCA), is a statistical analysis technique of obtaining the so called empirical eigenfunctions (EEFs) from the numerical or experimental data (Sirovich, 1987; Holmes *et al.*, 1996; Newman, 1996a). The basic idea of the KL expansion is to find those modes which represent the dominant character of the system.

The spatio-temporal dynamics can be separated into orthonormal spatial and temporal modes using the singular value decomposition (SVD)

$$y(x,t)=\sum_{i=1}^{\infty}\sigma_i\phi_i(x)\psi_i(t), \sigma_1\geq\sigma_2\geq\ldots\geq\sigma_\infty\,, \tag{2.26}$$

where singular values σ_i denote the importance of the modes, left singular functions $\phi_i(x)$ represent spatial modes, and right singular functions $\psi_i(t)$ are temporal modes. The KL method is actually implemented in different ways. Using the spatio-temporal data (snapshots), the KL method is transformed into an eigenfunction/eigenvalue problem of a spatial two-point correlation function (Sirovich, 1987)

$$\int_0^{\pi}C_x(x,\zeta)\phi_i(\zeta)d\zeta=\lambda_i\phi_i(x)\,, \tag{2.27}$$

where $C_x(x,\zeta)=\frac{1}{T}\int_0^T y(x,t)x(\zeta,t)dt$. This method is computationally efficient when the relevant number of snapshots is significantly larger than the dimension of the discretization. On the contrary, assuming that $\phi_i(x)=\sum_{t=1}^{T}\gamma_{it}y(x,t)$, then a smaller eigenvector/eigenvalue problem of a temporal two-point correlation function is used instead (Sirovich, 1987)

$$C_t\gamma_i=\lambda_i\gamma_i\,, \tag{2.28}$$

where $C_t(t,\tau)=\frac{1}{T}\int_0^{\pi}y(x,t)y(x,\tau)dx$.

The EEFs are found in an ordered manner based on the values of σ_i or λ_i ($\lambda_i=\sigma_i^2$), with the first EEF as the most dominant behavior, the second as the next dominant and so on. Usually, only the first few modes can capture the most important dynamics of the system, thus a small number of EEFs can be selected to yield a low-order model. Figure 2.8 shows the empirical eigenfunctions of Case 1.

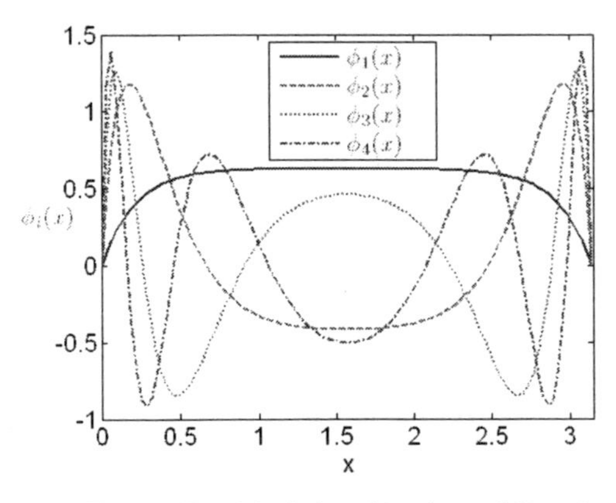

Fig. 2.8 Empirical eigenfunctions of Case 1

The *KL-Galerkin* method could be one of the most commonly used DPS modeling methods. It has been applied to system analysis, model reduction, numerical simulation of many complex distributed systems, e.g., fluid flow (Holmes *et al.*, 1996; Deane *et al.*, 1991; Park & Cho, 1996a), thermal process (Park & Cho, 1996b; Newman, 1996b; Banerjee, Cole & Jensen, 1998; Adomaitis, 2003), and diffusion-convection-reaction process (Armaou & Christofides, 2001c), etc. Many control applications can be carried out, e.g., control of the growth of thin films in a chemical vapor deposition reactor (Banks, Beeler, Kepler & Tran, 2002), control of a diffusion-reaction process (Armaou & Christofides, 2001a), control of a thin shell system (Banks, del Rosario & Smith, 2000), and optimization of diffusion-convection-reaction processes (Bendersky & Christofides, 2000).

The applications of *KL-Galerkin-AIM* method (Baker & Christofides, 2000; Aling *et al.*, 1997) and the *KL-collocation* method (Adomaitis, 1995; Baker & Christofides, 1999; Theodoropoulou *et al.*, 1998) are also studied for simulation, control and optimization of thermal processes.

Remark 2.3: ***Spectral Method and KL Method***

Compared with the spectral method, the KL method is applicable to a wider range of complex distributed systems, including those with irregular domain, nonlinear spatial operator and nonlinear boundary conditions. Because BFs of the KL method can provide an optimal linear representation of spatio-temporal data themselves (Holmes *et al.*, 1996; Dür, 1998), it may generate a lower-order and more accurate model than the FEM and the spectral method. However, its major drawback is that the KL method depends on cases and lacks a systematic solution for one class of systems. Thus, experiment settings such as the input signal, the time interval, the number of snapshots, the values of system parameters (Park & Cho, 1996b), and the initial conditions (Graham & Kevrekidis, 1996) have to be carefully chosen for an efficient application.

2.2.5 Comparison Studies of Spectral and KL Method

Spectral method stems from Fourier series expansion while Karhunen-Loève method comes from the idea of principle component analysis. There are some differences in the applications. Spectral method does not require any sensor, while KL method requires some sensors for learning basis functions. The number of basis functions resulting from KL method is fewer than or equal to the number of sensors used. Therefore the number of sensors determines the order of the model in KL method, while the number of eigenfunctions is equal to the model order in spectral method.

In order to compare the eigenfunctions and KL basis functions from a view of sensor number or model order, a modeling performance index is defined as *mean of absolute relative error* (MARE): $\frac{1}{NL}\sum_{i=1}^{N}\sum_{t=1}^{L}\left|\frac{y(x_i,t)-\hat{y}(x_i,t)}{y(x_i,t)}\right|$, where $y(x_i,t)$ and $\hat{y}(x_i,t)$ are true and estimated output at location x_i and time t, N and L are the number of spatial locations and the time length.

As shown in Figure 2.9, both spectral and KL methods are applicable for this quasi-linear parabolic system, where the horizontal axis is the model order for spectral method and the sensor number for KL method. The model order of KL method is fixed as 3. As the model order of spectral method or the sensor number of KL method increases, the modeling error becomes smaller. When the sensor number is too small, KL method will perform worse than spectral method. However, when the sensor number is large enough, KL method will be better than spectral method.

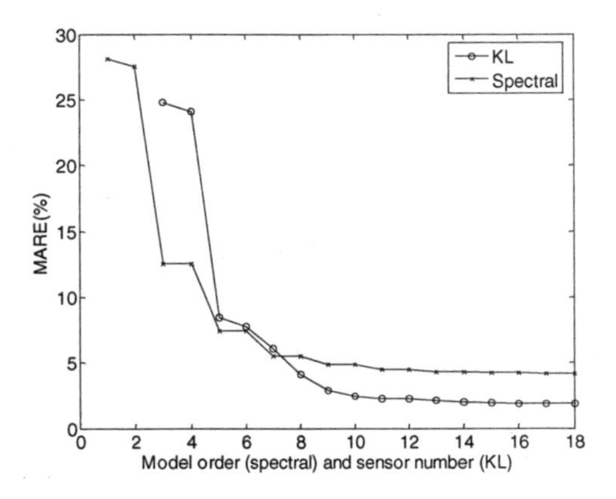

Fig. 2.9 KL and spectral method for Case 1

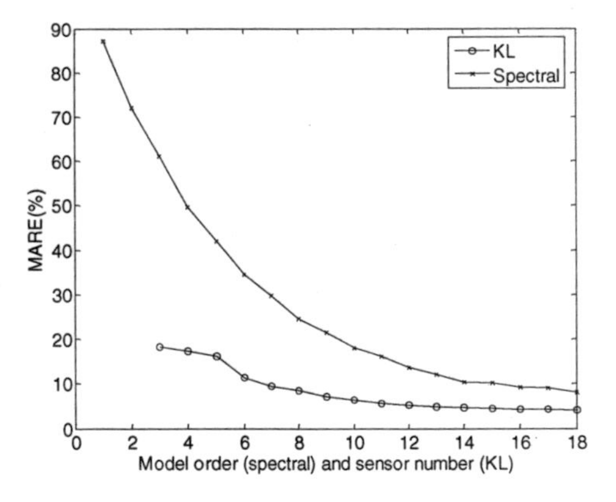

Fig. 2.10 KL and spectral method for Case 2

As shown in Figure 2.10, KL method is also efficient for this nonlinear parabolic system, while spectral method fails. This is because this system is very close to hyperbolic PDE, which does not show clearly separated eigenspectrum. In order to obtain the comparative accuracy, the number of sensors required in KL method is less than the model order in spectral method. In other words, KL basis functions may be more efficient than the eigenfunctions under some conditions, and that is obvious in the case of the non-separated eigenspectrum.

The limitation of spectral and KL method is that a PDE description of the process must be known. If the there is uncertainty in the process, the derived model will have a poorly approximation capability. In this case, the data-based modeling is often used to obtain an accurate model, which will be discussed in Section 2.4.4 - 2.4.6.

2.3 Grey-Box Modeling: Parameter Estimation for Partly Known DPS

All the previous model reduction methods require that the PDE description of the system is known. However, in many cases it is difficult to obtain an exact PDE description of the process from the physical and chemical laws only, so the data-based modeling must be employed to find the unknown DPS. In general, problems can be classified in two different cases: *parameter estimation* for known structure and *system identification* for unknown structure. In either case the model reduction methods introduced in Section 2.2 will play an important role.

This section will discuss parameter estimation problem for grey-box DPS modeling. When the PDE structure of the system is known and only some parameters are unknown, these parameters can be estimated from the experimental data (Banks & Kunisch, 1989). Once the parameters are determined, the PDE model reduction can be used for real applications, such as simulation, control, and optimization. Some earlier survey papers are given by Kubrusly (1977) and Polis & Goodson (1976).

2.3.1 FDM Based Estimation

After computing all the derivative terms in the PDE with the finite difference method, the unknown parameters can be estimated by minimizing the equation error (Coca & Billings, 2000; Müller & Timmer, 2004). It will lead to linear or nonlinear regression problem according to the type of PDE. Because it does not need to solve the PDE, this method is relatively simple. However, some data smoothing or de-noise techniques may be needed to reduce noise sensitivity. More sensors are usually used to obtain accurate parameter estimation if the parameters are temporally or spatially varying.

Since the predicted output can be computed from the dynamical PDE model using the finite difference method, and thus the system parameters can be found by minimizing the output error between the measurement and the prediction (Banks & Kunisch, 1989). For example, Kubrusly & Curtain, (1977) proposed a stochastic approximation algorithm for the spatially varying parameter estimation of the second-order linear parabolic PDE with the random input and noisy measurement. Uciński & Korbicz (1990) presented a recursive least-squares method to estimate spatially varying parameters of the parabolic DPS with noisy outputs. In addition, based on the ODE model derived from the *method of lines*, Vande Wouwer *et al.*, (2006) used a Levenberg-Marquardt algorithm to estimate constant parameters.

The comparison studies of the equation error and output error approaches are provided by Müller & Timmer (2004).

- The equation error approach is to first construct the left and right sides of the PDE in (2.1) from data, then to estimate the model by minimizing the equation residual given in (2.16).
- In the output error approach, the model is figured out by minimizing the error between the measured output and the predicted output as shown in Figure 2.11.

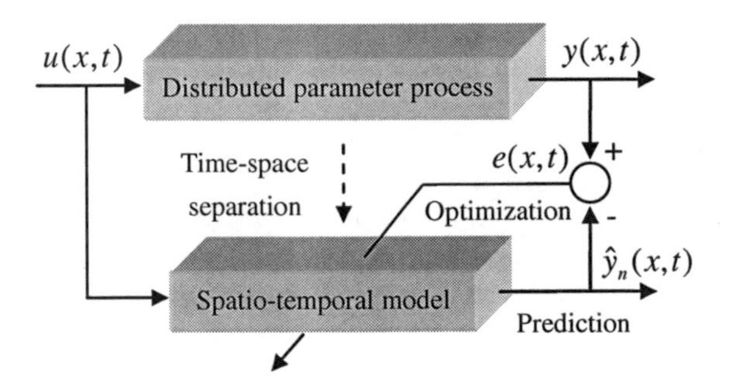

Fig. 2.11 Output error approach

Because the output error method needs to solve the PDE, its optimization computation may be complex. However, it does not need to estimate the derivative from the noisy data. The equation error and output error approaches can also be performed using *FEM*, *spectral* and *KL* methods.

2.3.2 FEM Based Estimation

Based on the *FEM-Galerkin* approximation, Rannacher & Vexler (2005) estimated the constant parameters for the elliptic PDE. With the finite spatial BF expansion of the parameters, Fernandez-Berdaguer, Santos & Sheen (1996) estimated the spatially varying parameters by minimizing the output error and regularization error with a quasi-linearization algorithm. The regularization error is formed from the output error by adding a penalty for model complexity which will help to solve an ill-posed problem or to prevent the overfitting. Based on the finite element parameterization of the parameters, Carotenuto & Raiconi (1980) estimated the spatially varying parameters of the linear parabolic DPS using nonlinear optimization techniques. Uciński & Korbicz (1990) proposed a recursive least-squares method to estimate spatially varying parameters of the parabolic DPS under noisy outputs.

Based on the *FEM-spline-Galerkin* approximation, Banks, Crowley & Kunisch (1983) studied the parameters estimation for the linear and nonlinear second-order parabolic and hyperbolic PDE systems. With the piecewise linear splines approximation of the parameters, Banks & Daniel Lamm (1985) presented the spatially and temporally varying parameters estimation for the linear parabolic system. With the Legendre polynomial or splines function approximation of the parameters, Banks, Reich & Rosen, (1991) estimated unknown spatially and/or temporally varying parameters of the linear or nonlinear DPS. With the finite-dimensional parameterization, Demetriou & Rosen (1994) proposed an adaptive parameter estimation algorithm for a class of second-order linear distributed systems with spatially varying parameters.

2.3.3 Spectral Based Estimation

Each term in the PDE is reconstructed from the data using the finite-dimensional expansion onto a set of spatial and temporal orthogonal functions, and then double integrating the PDE with respect to time and space will convert the equation error minimization problem to the problem of the algebraic equation (AE), which can be easily solved using linear regression method. This method is simple but requires the PDE to be a linear or a special nonlinear form for the AE conversion. Furthermore, many sensors may be needed due to its spatial integration. Under this approach, various orthogonal functions, e.g., Fourier series (Mohan & Datta, 1989), Chebyshev series (Horng, Chou & Tsai, 1986), Walsh-functions (Paraskevopoulos & Bounas, 1978), Laguerre polynomials (Ranganathan, Jha & Rajamani, 1984, 1986), Taylor series (Chung & Sun, 1988), general orthogonal polynomials (Lee & Chang, 1986) and block-pulse functions (Mohan & Datta, 1991), have been used for the parameter estimation.

Based on the *EF-Galerkin* method, an identification algorithm is performed using a Newton-like method for unknown constant diffusivities in the diffusion equation described by the parabolic PDE (Omatu & Matumoto, 1991). Using Lagrange interpolation polynomials as BFs of the *collocation* method, Ding, Gustafsson & Johansson (2007) studied the constant parameters estimation of a continuous paper pulp digester described by two linear hyperbolic PDE systems.

2.3.4 KL Based Estimation

Based on the *KL-Galerkin* approximation, Park, Kim & Cho (1998) presented the constant parameters estimation of flow reactors described by Navier-Stokes equation. Ghosh, Ravi Kumar & Kulkarni (2001) proposed a multiple shooting algorithm to estimate unknown initial conditions and constant parameters for a coupled map lattice (CML) system and a reaction-diffusion system.

2.4 Black-Box Modeling: System Identification for Unknown DPS

For the DPS with unknown structure which widely exist in real life applications, the black-box identification has to be used. More sensors are often required in spatial locations to collect enough information. Modeling becomes extremely difficult because both the structure and parameters need to be figured out. There are many system identification methods for LPS. Only a few studies discuss system identification of DPS.

2.4.1 Green's Function Based Identification

If the analytical Green's function is not available, it can be estimated from the input-output data. For example, based on the singular function expansion of Green's function, a time-invariant Green's function model can be estimated using SVD method (Gay & Ray, 1995). This approach yielded accurate low-order solutions for linear invariant non-self-adjoint systems. A disadvantage of this approach is the time-invariance assumption. Obviously for time-varying systems, this approach is limited. To avoid the assumption, a time-varying Green's function identification method is proposed by combining the characteristics of singular value decomposition and the Karhunen-Loève expansion (SVD-KL) (Zheng, Hoo & Piovoso, 2002; Zheng & Hoo, 2002; Zheng & Hoo, 2004). The Green's function can also be estimated using other methods (Doumanidis & Fourligkas, 2001). Because the Green's function model uses one single kernel, it may only be able to approximate the linear DPS or the nonlinear DPS at the given working condition. Therefore how to extend the kernel idea to the nonlinear DPS is an important problem, which will be discussed in Chapter 6.

2.4.2 FDM Based Identification

A basic idea is that after a candidate set of spatial and temporal derivatives are estimated from the data using finite difference, the functions representing the model can be determined by minimizing an equation error criterion using the optimization techniques. Voss, Bünner & Abel (1998) presented alternating conditional expectation algorithm for parameter estimation under a pre-selected model structure. Bär, Hegger & Kantz (1999) used singular value decomposition or backward elimination method for the model selection.

To avoid computing temporal derivatives with finite difference for noisy data, the PDE system turns to unknown algebraic equations using implicit Adams integration over time (Guo & Billings, 2006). Then the system structure and parameters can be estimated by an orthogonal least-squares algorithm, in which the PDE operators are expanded using polynomials as BFs, and the spatial derivatives are estimated by finite difference methods. To further reduce noise sensitivity, the spatial derivatives can be estimated using a B-spline functions-based multi-resolution analysis instead of finite difference methods (Guo, Billings & Wei, 2006).

The above method can obtain an ODE or PDE model from the data. However, the ODE model will be high-order. When the PDE model is obtained, model reduction is still needed for the practical applications. Moreover, it may lead to a complicated PDE model, whose reduced model may not be suitable for practical process control.

The time-space discretization of the PDE using *FDM* can lead to the difference equation (DE), which can be considered as so called lattice dynamical system (LDS), thus a class of DE model identification methods are based on the identification of lattice dynamical system (Parlitz & Merkwirth, 2000; Mandelj, Grabec & Govekar, 2001; Guo & Billings, 2007). The key feature is that the model of the system is usually unchanged at different spatial locations (except the boundary) in Figure 2.12, and the dynamics of the node (black node) is only determined by its neighbor regions (white nodes). For example, Mandelj, Grabec & Govekar (2001, 2004) and Abel (2004) used nonparametric statistical identification method, and Coca & Billings (2001), Billings & Coca (2002) and Guo & Billings (2007) estimated the parametric nonlinear autoregressive with exogenous input (NARX) model using orthogonal forward regression (OFR) algorithm.

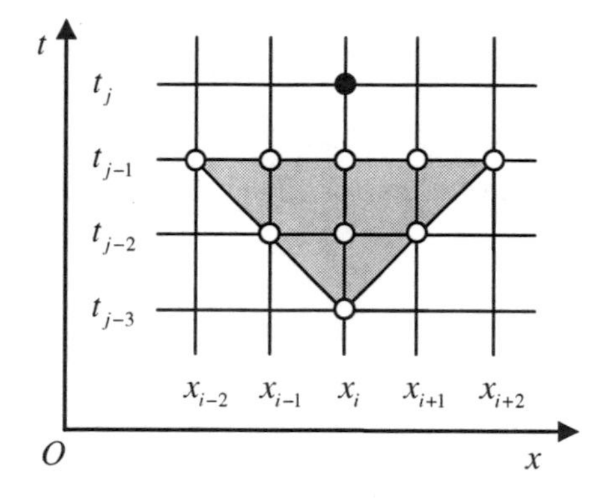

Fig. 2.12 Geometric interpretation of FDM based identification

A good model can be obtained if the small regions are properly determined. Thus one task is to determine the proper neighborhood. The heuristic or pre-specified approaches (Parlitz & Merkwirth, 2000; Mandelj, Grabec & Govekar, 2001) as well as the theoretical analysis (Guo & Billings, 2007) have been proposed. Each local model at the node represents a state of the system, thus the model may be of a high-dimension since its dimension is determined by the number of spatial discretization points.

If the *method of lines* is used with only spatial discretization, the resulting ODE model can be estimated using traditional ODE identification techniques such as neural networks (Gonzalez-Garcia, Rico-Martinez & Kevrekidis, 1998). It has similar weakness as the LDS modeling, i.e., the dimension will be high-order and the model has complex nonlinear structure.

2.4.3 FEM Based Identification

After choosing the proper BFs, the corresponding unknown temporal model can be estimated using traditional system identification techniques. The methodology is similar to that shown in Figure 2.3, where the temporal model is identified from the data instead of derivation from the PDE description.

The basic procedure can be expressed as follows. Suppose we have a set of sampled process input $u(x_i,t)$ ($i=1,...,M$) and output $y(x_i,t)$ ($i=1,...,N$). Firstly the basis function expansion

$$u(x,t) \approx \sum_{i=1}^{m} u_i(t)\varphi_i(x) , \tag{2.29}$$

$$y(x,t) \approx \sum_{i=1}^{n} y_i(t)\phi_i(x) , \tag{2.30}$$

is used for the time/space separation and dimension reduction. Secondly the modeling problem is to estimate a finite order temporal model

$$y(t+1) = F(y(t),u(t)) , \tag{2.31}$$

from $u(t) = [u_1(t),\cdots,u_m(t)]^T \in \mathbb{R}^m$ and $y(t) = [y_1(t),\cdots,y_n(t)]^T \in \mathbb{R}^n$, where $F:\mathbb{R}^n \times \mathbb{R}^m \to \mathbb{R}^n$ is a nonlinear function. Finally the output equation

$$y_n(z,t) = \sum_{i=1}^{n} y_i(t)\phi_i(z) , \tag{2.32}$$

is used for the time/space synthesis.

Various lumped system identification techniques can be used to identify the unknown function F, and the neural network is often used among them. By working with different selections of BFs, FEM, spectral and KL based identification can be formulated.

Coca & Billings (2002) proposed a FEM based identification method. With local basis functions chosen as B-spline functions, the NARX model is identified to recover the system dynamics using the OFR algorithm for model selection and parameter estimation, with applications to a linear diffusion system and a nonlinear reaction-diffusion system. However, the model may be high-order for a satisfactory modeling accuracy because of local basis functions. In order to obtain a low-order model, the spectral and KL based identification may be two good approaches.

2.4.4 Spectral Based Identification

Based on the eigenfunctions expansion and artificial neural network approximation, a neural spectral method is proposed as shown in Figure 2.13. Of course, other global basis functions can also be used though they may not be optimal in sense that the model order could be high. To obtain the eigenfunctions, the nominal linear model of the system should be known. Neural spectral method usually needs more sensors, and the number of sensors at least equals or exceeds the number of eigenfunctions required for modeling.

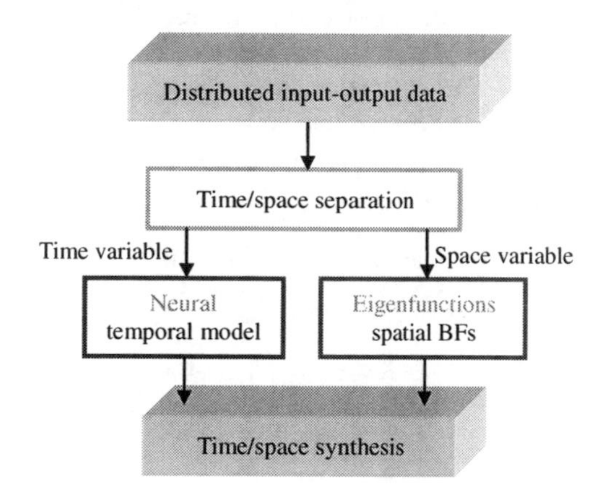

Fig. 2.13 Neural spectral method

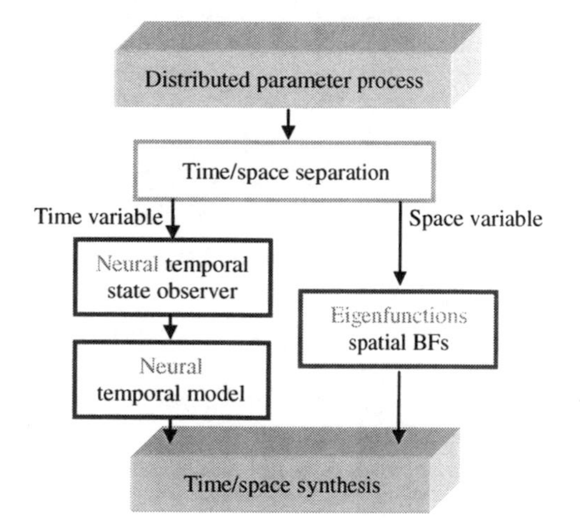

Fig. 2.14 Neural observer spectral method

A neural observer spectral method is proposed as shown in Figure 2.14 (Deng, Li & Chen, 2005), which usually needs a smaller number of sensors. The basic idea is that by utilizing a priori knowledge about the nominal model, a neural observer can be developed to estimate the higher-dimensional states from a few sensors. Therefore, a better modeling performance is obtained with fewer sensors. This method is useful for a class of partly known PDE considered with unknown parameters and nonlinearities. A simple model-based control is further developed for a class of curing thermal processes (Li, Deng & Zhong, 2004). However, as mentioned in Section 1.2, neural observer spectral method has some limitations: (1) The model order is relatively high. (2) The model has a complex nonlinear structure which is not easy for control design. (3) It is not easy to apply to other industrial processes because the intelligent observer design is very dependent on a priori knowledge of the process. This motivates us to develop simple low-order nonlinear models with less process knowledge required.

2.4.5 KL Based Identification

The neural networks are popularly integrated with the KL method to identify unknown DPS (Zhou, Liu, Dai & Yuan, 1996), as shown in Figure 2.15.

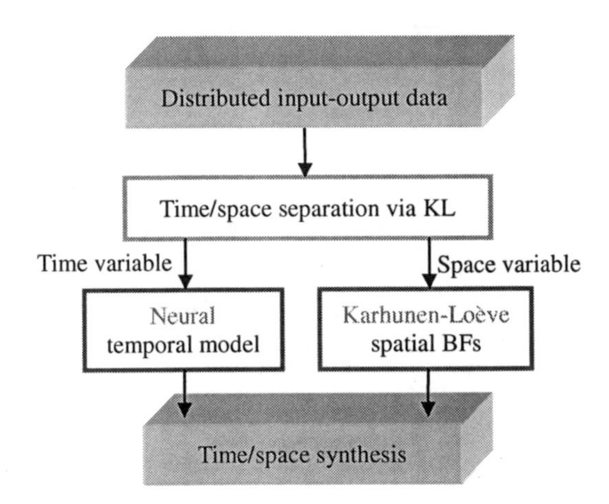

Fig. 2.15 Neural KL method

Applications include: analysis of Kuramoto-Sivashinsky and Navier-Stokes equations (Smaoui & Al-Enezi, 2004), modeling and intelligent control of transitional flows (Sahan *et al.*, 1997), modeling of a heat convection-conduction process with the known mechanistic part and unknown nonlinearity (Romijn *et al.*, 2008). In all these work, the neural network is used for learning the nonlinearity. However, design of the neural system is not very systematic. Recently, the modeling and predictive control with applications to a tubular reactor is studied by using a fuzzy partition method for network design (Aggelogiannaki & Sarimveis, 2008). Though neural KL method may achieve a low-order model, most neural models are very complex which are actually not suitable for practical control. The approach using KL method and subspace identification is also reported for bilinear model identification and receding horizon control of a thin-film growth process (Varshney & Armaou, 2008).

2.4.6 Comparison Studies of Neural Spectral and Neural KL Method

For the neural KL method, more efficient KL basis functions are used as spatial basis function instead of the eigenfunctions in the neural spectral method. A fact is that the eigenfunction is calculated from the system, and therefore the neural spectral method requires the linear nominal model to be known, while the neural KL method does not require any a priori knowledge of the system.

As shown in Figure 2.16, both neural spectral and neural KL models can give a better performance than the nominal spectral model for Case 1 when enough sensors are used, because the nonlinear uncertainties in the system can be compensated by neural network learning using the sensor measurements. The horizontal axis is the model order for the nominal spectral method and the sensor number for neural spectral and neural KL method. The model order of neural KL method is fixed at three, where the model order of the neural spectral is set to be the number of sensors used. It also shows that the neural KL model even with smaller number of basis functions is better than the neural spectral method because of the more efficient Karhunen-Loève method.

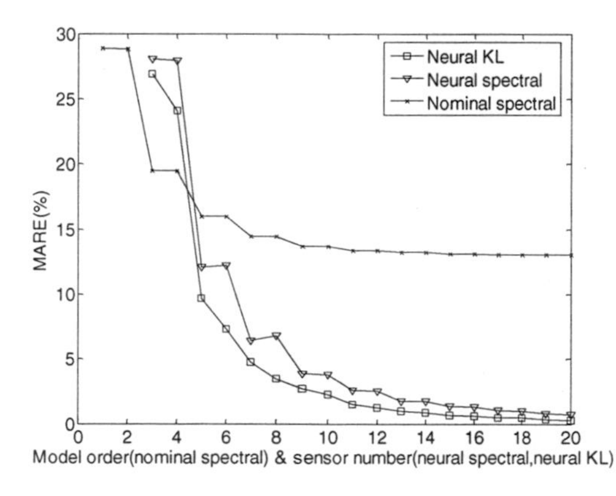

Fig. 2.16 Neural spectral and neural KL methods for Case 1

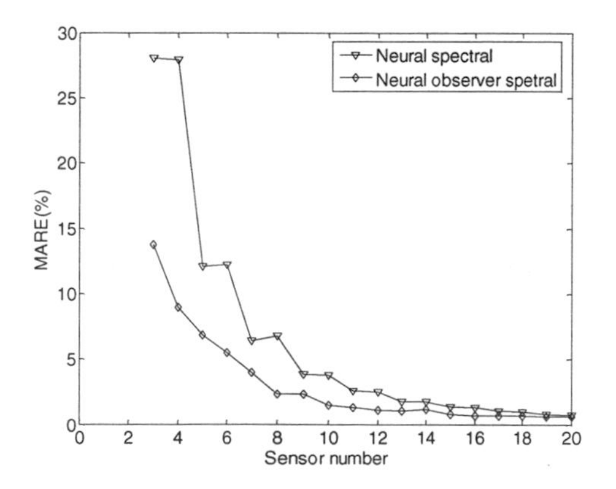

Fig. 2.17 Neural spectral and neural observer spectral methods for Case 1

Both the neural spectral and the neural KL method need enough sensors for a good performance. As shown in Figure 2.17, for the neural observer spectral method, the required sensors are less than those of the neural spectral method.

As shown in Figure 2.18 and Figure 2.19, the neural spectral model performs much worse than the neural KL model for Case 2, where the model order of the neural spectral is equal to the sensor number while the model order of the neural KL is set to be four. It can be said the neural spectral method does not work. The neural observer spectral method also fails in this case. This is because the neural spectral and the neural observer spectral method are only suitable for the case that has significant separation of eigenspectrum. Alternately, the neural KL model is applicable to a wider range of systems.

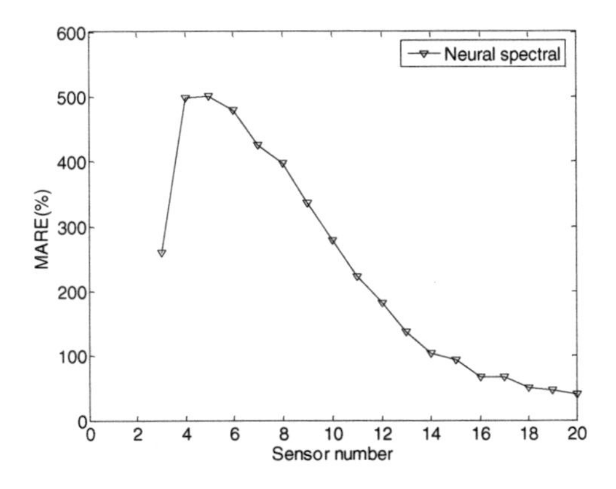

Fig. 2.18 Neural spectral method for Case 2

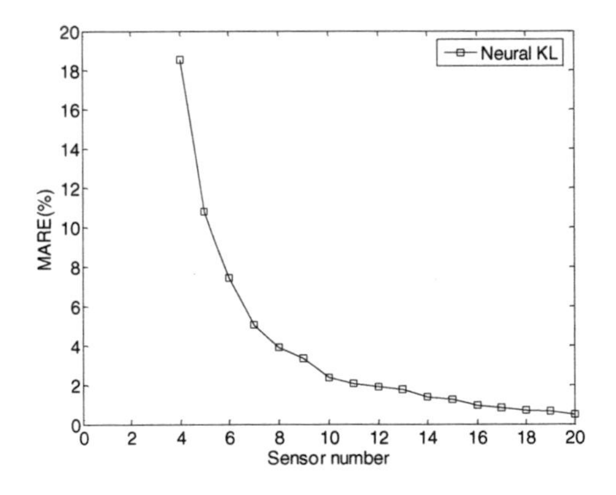

Fig. 2.19 Neural KL method for Case 2

The neural network models developed above can approximate many nonlinear systems. However, it is not easy to be used for control design because of their inherent nonlinear property. In practice, a model with simple or special nonlinear structure is often needed. Because the KL method can achieve a lower-order model than the spectral method and is suitable for a wider range of DPS, so it will be used in the data-based modeling development.

2.5 Concluding Remarks

This chapter presents an overview on advances in the DPS modeling. The DPS modeling problem can be classified into three issues: *model reduction* for known DPS, *parameter estimation* for DPS with unknown parameters and *system identification* for DPS with unknown structure. Different methods have been classified into several categories according to their methodology. The underlying fundamental ideas and their strength and weakness are also presented. For numerical implementation and applications, the infinite-dimensional problem must be transformed to a finite-dimensional one, where the *model reduction* is fundamental.

The model reduction and parameter estimation of DPS are relatively mature. However, there are some problems in the system identification of DPS, which will be studied in this book.

- For the DPS system identification in Section 2.4.2 - 2.4.5, most of models use neural network methods. They have complex and general nonlinear structure, which may be difficult for control design. In the LPS modeling, the block-oriented nonlinear model (e.g., Hammerstein and Wiener model) has been widely used because of their simple structures, abilities to approximate a large class of nonlinear processes and efficient control schemes (e.g., Narendra & Gallman, 1966; Stoica & Söderström, 1982; Bai, 1998; Zhu, 2000; Gómez & Baeyens, 2004; Westwick & Verhaegen, 1996; Hagenblad & Ljung, 2000). They consist of the interconnection of linear time invariant (LTI) systems and static nonlinearities. To best of our knowledge, the block-oriented nonlinear models are only studied for LPS. The extension of blocked-oriented nonlinear models to the spatio-temporal system is very useful for modeling the nonlinear DPS. Thus in Chapter 3 we study the spatio-temporal modeling for the Wiener distributed parameter system. In Chapters 4 and 5, the spatio-temporal modeling will be studied for the Hammerstein distributed parameter system.
- Though the Green's function model in Section 2.4.1 is very suitable for control design because of its linear structure, however, it is only a linear approximation for a nonlinear DPS. How to identify a nonlinear DPS using the kernel is an important problem. Thus, the kernel based spatio-temporal modeling is presented for the nonlinear Hammerstein distributed parameter system in Chapter 5. Volterra model is widely used for modeling nonlinear LPS (Boyd & Chua, 1985; Schetzen, 1980; Rugh, 1981; Doyle III *et al.*, 1995; Maner *et al.*, 1996; Parker *et al.*, 2001). However, it consists of a series of temporal kernels, the extension to the spatio-temporal system is very necessary for modeling the nonlinear DPS. Thus Volterra kernel based spatio-temporal modeling is presented in Chapter 6.

- For the KL based neural modeling in Section 2.4.5, the use of KL for dimension reduction is only a linear approximation for a nonlinear DPS. In the field of machine learning, the nonlinear principal component analysis (NL-PCA) has been widely used for nonlinear dimension reduction of high-dimensional data or known systems (e.g., Dong & McAvoy, 1996; Kramer, 1991; Hsieh, 2001; Kirby & Miranda, 1994; Smaoui, 2004; Webb, 1996; Wilson, Irwin & Lightbody, 1999). The application of nonlinear dimension reduction methods to the nonlinear DPS will improve the KL based modeling performance. Thus in Chapter 7, the nonlinear dimension reduction based neural modeling will be discussed for the nonlinear DPS.

References

1. Abel, M.: Nonparametric modeling and spatiotemporal dynamical systems. International Journal of Bifurcation and Chaos 14(6), 2027–2039 (2004)
2. Adomaitis, R.A.: RTCVD model reduction: A collocation on empirical eigenfunctions approach. Technical Report T.R.95-64, University of Maryland, College Park, Maryland (1995)
3. Adomaitis, R.A.: A reduced-basis discretization method for chemical vapor deposition reactor simulation. Mathematical and Computer Modeling 38(1-2), 59–175 (2003)
4. Adrover, A., Continillo, G., Crescitelli, S., Giona, M., Russo, L.: Construction of approximate inertial manifold by decimation of collocation equations of distributed parameter systems. Computers and Chemical Engineering 26(1), 113–123 (2002)
5. Adrover, A., Continillo, G., Crescitelli, S., Giona, M., Russo, L.: Wavelet-like collocation method for finite-dimensional reduction of distributed systems. Computers and Chemical Engineering 24(12), 2687–2703 (2000)
6. Aggelogiannaki, E., Sarimveis, H.: Nonlinear model predictive control for distributed parameter systems using data driven artificial neural network models. Computers and Chemical Engineering 32(6), 1225–1237 (2008)
7. Aling, H., Banerjee, S., Bangia, A.K., Cole, V., Ebert, J., Emani-Naeini, A., Jensen, K.F., Kevrekidis, I.G., Shvartsman, S.: Nonlinear model reduction for simulation and control of rapid thermal processing. In: Proceedings of the 1977 American Control Conference, Albuquerque, New Mexico, pp. 2233–2238 (1997)
8. Armaou, A., Christofides, P.D.: Nonlinear feedback control of parabolic partial differential equation systems with time-dependent spatial domains. Journal of Mathematical Analysis & Applications 239(1), 124–157 (1999)
9. Armaou, A., Christofides, P.D.: Finite-dimensional control of nonlinear parabolic PDE systems with time-dependent spatial domains using empirical eigenfunctions. International Journal of Applied Mathematics and Computer Science 11(2), 287–317 (2001a)
10. Armaou, A., Christofides, P.D.: Robust control of parabolic PDE systems with time-dependent spatial domains. Automatica 37(1), 61–69 (2001b)
11. Armaou, A., Christofides, P.D.: Computation of empirical eigenfunctions and order reduction for nonlinear parabolic PDE systems with time-dependent spatial domains. Nonlinear Analysis 47(4), 2869–2874 (2001c)
12. Bai, E.W.: An optimal two-stage identification algorithm for Hammerstein-Wiener nonlinear systems. Automatica 34(3), 333–338 (1998)

13. Baker, J., Christofides, P.D.: Output feedback control of parabolic PDE systems with nonlinear spatial differential operators. Industrial & Engineering Chemistry Research 38(11), 4372–4380 (1999)
14. Baker, J., Christofides, P.D.: Finite-dimensional approximation and control of non-linear parabolic PDE systems. International Journal of Control 73(5), 439–456 (2000)
15. Balas, M.J.: The Galerkin method and feedback control of linear distributed parameter systems. Journal of Mathematical Analysis and Applications 91(2), 527–546 (1983)
16. Balas, M.J.: Finite dimensional control of distributed parameter systems by Galerkin approximation of infinite dimensional controller. Journal of Mathematical Analysis and Application 114, 17–36 (1986)
17. Banerjee, S., Cole, J.V., Jensen, K.F.: Nonlinear model reduction strategies for rapid thermal processing systems. IEEE Transactions on Semiconductor Manufacturing 11(2), 266–275 (1998)
18. Banks, H.T., Daniel Lamm, P.K.: Estimation of variable coefficients in parabolic distributed systems. IEEE Transactions of Automatic Control 30(4), 386–398 (1985)
19. Banks, H.T., Kunisch, K.: Estimation techniques for distributed parameter systems. Birkhauser, Boston (1989)
20. Banks, H.T., Beeler, R.C., Kepler, G.M., Tran, H.T.: Reduced-order model modeling and control of thin film growth in a HPCVD reactor. SIAM Journal on Applied Mathematics 62(4), 1251–1280 (2002)
21. Banks, H.T., Crowley, J.M., Kunisch, K.: Cubic spline approximation techniques for parameter estimation in distributed systems. IEEE Transactions on Automatic Control 28(7), 773–786 (1983)
22. Banks, H.T., del Rosario, R.C.H., Smith, R.C.: Reduced-order model feedback control design: Numerical implementation in a thin shell model. IEEE transactions on Automatic Control 45(7), 1312–1324 (2000)
23. Banks, H.T., Reich, S., Rosen, I.G.: Galerkin approximation for inverse problems for nonautonomous nonlinear distributed systems. Applied Mathematics and Optimization 24(3), 233–256 (1991)
24. Bär, M., Hegger, R., Kantz, H.: Fitting partial differential equations to space-time dynamics. Physical Review E 59(1), 337–342 (1999)
25. Bendersky, E., Christofides, P.D.: Optimization of transport-reaction processes using nonlinear model reduction. Chemical Engineering Science 55(19), 4349–4366 (2000)
26. Billings, S.A., Coca, D.: Identification of coupled map lattice models of deterministic distributed parameter systems. International Journal of Systems Science 33(8), 623–634 (2002)
27. Boyd, J.P.: Chebyshev and Fourier spectral methods, 2nd edn. Dover Publications, New York (2000)
28. Boyd, S., Chua, L.O.: Fading memory and the problem of approximating nonlinear operators with Volterra series. IEEE Transactions on Circuits and Systems 32(11), 1150–1161 (1985)
29. Brenner, S.C., Ridgway Scott, L.: The mathematical theory of finite element methods. Springer, New York (1994)
30. Butkovskiy, A.G.: Green's functions and transfer functions handbook, 1st edn. Ellis Horwood, Chichester (1982)
31. Canuto, C., et al.: Spectral methods in fluid dynamics. Springer, New York (1988)

32. Carotenuto, L., Raiconi, G.: Identifiability and identification of a Galerkin approximation for a class of distributed parameter-systems. International Journal of Systems Science 11(9), 1035–1049 (1980)
33. Christofides, P.D.: Robust control of parabolic PDE systems. Chemical Engineering Science 53(16), 2949–2965 (1998)
34. Christofides, P.D.: Nonlinear and robust control of PDE systems: Methods and applications to transport-reaction processes. Birkhäuser, Boston (2001b)
35. Christofides, P.D., Baker, J.: Robust output feedback control of quasi-linear parabolic PDE systems. Systems & Control Letters 36(5), 307–316 (1999)
36. Christofides, P.D., Daoutidis, P.: Finite-dimensional control of parabolic PDE systems using approximate inertial manifolds. Journal of Mathematical Analysis and Applications 216(2), 398–420 (1997)
37. Chung, H.Y., Sun, Y.Y.: Parameter identification of linear distributed systems via Taylor operational matrix. IEEE Transactions on Industrial Electronics 35(3), 413–416 (1988)
38. Coca, D., Billings, S.A.: Direct parameter identification of distributed parameter systems. International Journal of Systems Science 31(1), 11–17 (2000)
39. Coca, D., Billings, S.A.: Identification of coupled map lattice models of complex spatio-temporal patterns. Physics Letters A 287(1-2), 65–73 (2001)
40. Coca, D., Billings, S.A.: Identification of finite dimensional models of infinite dimensional dynamical systems. Automatica 38(11), 1851–1865 (2002)
41. Cruz, P., Mendes, A., Magalhaes, F.D.: Using wavelets for solving PDEs: an adaptive collocation method. Chemical Engineering Science 56(10), 3305–3309 (2001)
42. Deane, A.E., Kevrekidis, I.G., Karniadakis, G.E., Orszag, S.A.: Low dimensional models for complex geometry flows: Application to grooved channels and circular cylinders. Fluids Physics A 3(10), 2337–2354 (1991)
43. Demetriou, M.A., Rosen, I.G.: Adaptive identification of second-order distributed parameter systems. Inverse Problems 10(2), 261–294 (1994)
44. Deng, H., Li, H.-X., Chen, G.: Spectral-approximation-based intelligent modeling for distributed thermal processes. IEEE Transactions on Control Systems Technology 13(5), 686–700 (2005)
45. Ding, L., Gustafsson, T., Johansson, A.: Model parameter estimation of simplified linear models for a continuous paper pulp digester. Journal of Process Control 17(2), 115–127 (2007)
46. Dochain, D., Babary, J.P., Tali-Maamar, N.: Modelling and adaptive control of nonlinear distributed parameter bioreactors via orthogonal collocation. Automatica 28(5), 873–883 (1992)
47. Dong, D., McAvoy, T.J.: Nonlinear principal component analysis-based on principal curves and neural networks. Computers and Chemical Engineering 20(1), 65–78 (1996)
48. Doumanidis, C.C., Fourligkas, N.: Temperature distribution control in scanned thermal processing of thin circular parts. IEEE Transactions on Control Systems Technology 9(5), 708–717 (2001)
49. Doyle III, F.J., Ogunnaike, B.A., Pearson, R.K.: Nonlinear model-based control using second-order Volterra Models. Automatica 31(5), 697–714 (1995)
50. Dubljevic, S., Christofides, P.D., Kevrekidis, I.G.: Distributed nonlinear control of diffusion-reaction processes. International Journal of Robust and Nonlinear Control 14(2), 133–156 (2004)

51. Dür, A.: On the optimality of the discrete Karhunen-Loève expansion. SIAM Journal on Control and Optimization 36(6), 1937–1939 (1998)
52. El-Farra, N.H., Christofides, P.D.: Coordinating feedback and switching for control of spatially distributed processes. Computers and Chemical Engineering 28(1-2), 111–128 (2004)
53. El-Farra, N.H., Armaou, A., Christofides, P.D.: Analysis and control of parabolic PDE systems with input constraints. Automatica 39(4), 715–725 (2003)
54. Fernandez-Berdaguer, E.M., Santos, J.E., Sheen, D.: An iterative procedure for estimation of variable coefficients in a hyperbolic system. Applied Mathematics and Computation 76(2-3), 213–250 (1996)
55. Fletcher, C.A.J.: Computational Galerkin methods, 1st edn. Springer, New York (1984)
56. Foias, C., Temam, R.: Algebraic approximation of attractors: The finite dimensional case. Physica D 32(2), 163–182 (1988)
57. Foias, C., Jolly, M.S., Kevrikidis, I.G., Sell, G.R., Titi, E.S.: On the computation of inertial manifolds. Physics Letters A 131(7-8), 433–436 (1988)
58. Gay, D.H., Ray, W.H.: Identification and control of distributed parameter systems by means of the singular value decomposition. Chemical Engineering Science 50(10), 1519–1539 (1995)
59. Ghosh, A., Ravi Kumar, V., Kulkarni, B.D.: Parameter estimation in spatially extended systems: The Karhunen-Loève and Galerkin multiple shooting approach. Physical Review E 64(5), 56222 (2001)
60. Gómez, J.C., Baeyens, E.: Identification of block-oriented nonlinear systems using orthonormal bases. Journal of Process Control 14(6), 685–697 (2004)
61. Gonzalez-Garcia, R., Rico-Martinez, R., Kevredidis, I.G.: Identification of distributed parameter systems: A neural net based approach. Computers and Chemical Engineering 22, S965–S968 (1998)
62. Graham, M.D., Kevrekidis, I.G.: Alternative approaches to the Karhunen-Loève decomposition for model reduction and data analysis. Computers and Chemical Engineering 20(5), 495–506 (1996)
63. Guo, L.Z., Billings, S.A.: Identification of partial differential equation models for continuous spatio-temporal dynamical systems. IEEE Transactions on Circuits and Systems - II: Express Briefs 53(8), 657–661 (2006)
64. Guo, L.Z., Billings, S.A.: Sate-space reconstruction and spatio-temporal prediction of lattice dynamical systems. IEEE Transactions on Automatic Control 52(4), 622–632 (2007)
65. Guo, L.Z., Billings, S.A., Wei, H.L.: Estimation of spatial derivatives and identification of continuous spatio-temporal dynamical systems. International Journal of Control 79(9), 1118–1135 (2006)
66. Hagenblad, A., Ljung, L.: Maximum likelihood identification of Wiener models with a linear regression initialization. In: Proceedings of the 37th IEEE Conference Decision & Control, Tampa, Fourida, USA, pp. 712–713 (1998)
67. Höllig, K.: Finite element methods with B-splines. Society Industrial and Applied Mathematics, Philadelphia (2003)
68. Holmes, P., Lumley, J.L., Berkooz, G.: Turbulence, coherent structures, dynamical systems, and symmetry. Cambridge University Press, New York (1996)
69. Hoo, K.A., Zheng, D.: Low-order control-relevant models for a class of distributed parameter systems. Chemical Engineering Science 56(23), 6683–6710 (2001)

70. Horng, I.-R., Chou, J.-H., Tsai, C.-H.: Analysis and identification of linear distributed systems via Chebyshev series. International Journal of Systems Science 17(7), 1089–1095 (1986)
71. Hsieh, W.W.: Nonlinear principal component analysis by neural networks. Tellus Series A - Dynamic Meteorology and Oceanography 53(5), 599–615 (2001)
72. Kirby, M., Miranda, R.: The nonlinear reduction of high-dimensional dynamical systems via neural networks. Physical Review Letter 72(12), 1822–1825 (1994)
73. Ko, J., Kurdila, A.J., Pilant, M.S.: A class of finite element methods based on orthonormal compactly supported wavelets. Computational Mechanics 16(4), 235–244 (1995)
74. Kramer, M.A.: Nonlinear principal component analysis using autoassociative neural networks. AIChE Journal 37(2), 233–243 (1991)
75. Kubrusly, C.S.: Distributed parameter system identification - a survey. International Journal of Control 26(4), 509–535 (1977)
76. Kubrusly, C.S., Curtain, R.F.: Identification of noisy distributed parameter systems using stochastic approximation. International Journal of Control 25(3), 441–455 (1977)
77. Lee, T.-T., Chang, Y.-F.: Analysis and identification of linear distributed systems via double general orthogonal polynomials. International Journal of Control 44(2), 395–405 (1986)
78. Lefèvre, L., Dochain, D., Feyo de Azevedo, S., Magnus, A.: Optimal selection of orthogonal polynomials applied to the integration of chemical reactor equations by collocation methods. Computers and Chemical Engineering 24(12), 2571–2588 (2000)
79. Li, H.-X., Deng, H., Zhong, J.: Model-based integration of control and supervision for one kind of curing process. IEEE Transactions on Electronics Packaging Manufacturing 27(3), 177–186 (2004)
80. Mahadevan, N., Hoo, K.A.: Wavelet-based model reduction of distributed parameter systems. Chemical Engineering Science 55(19), 4271–4290 (2000)
81. Mandelj, S., Grabec, I., Govekar, E.: Statistical approach to modeling of spatiotemporal dynamics. International Journal of Bifurcation and Chaos 11(11), 2731–2738 (2001)
82. Mandelj, S., Grabec, I., Govekar, E.: Nonparametric statistical modeling approach of spatiotemporal dynamics based on recorded data. International Journal of Bifurcation and Chaos 14(6), 2011–2025 (2004)
83. Maner, B.R., Doyle III, F.J., Ogunnaike, B.A., Pearson, R.K.: Nonlinear model predictive control of a simulated multivariable polymerization reactor using second-order Volterra models. Automatica 32(9), 1285–1301 (1996)
84. Marion, M., Temam, R.: Nonlinear Galerkin methods: The finite elements case. Numerische Mathematik 57(1), 205–226 (1990)
85. Mitchell, A.R., Griffiths, D.F.: The finite difference method in partial differential equations. Wiley, Chichester (1980)
86. Mohan, B.M., Datta, K.B.: Identification via Fourier series for a class of lumped and distributed parameter systems. IEEE Transactions on Circuits and Systems 36(11), 1454–1458 (1989)
87. Mohan, B.M., Datta, K.B.: Linear time-invariant distributed parameter system identification via orthogonal functions. Automatica 27(2), 409–412 (1991)
88. Müller, T.G., Timmer, J.: Parameter identification techniques for partial differential equations. International Journal of Bifurcation and Chaos 14(6), 2053–2060 (2004)

89. Narendra, K., Gallman, P.: An iterative method for the identification of nonlinear systems using a Hammerstein model. IEEE Transactions on Automatic Control 11(3), 546–550 (1966)
90. Newman, A.J.: Model reduction via the Karhunen-Loève expansion part I: An exposition. Technical Report T.R.96-32, University of Maryland, College Park, Maryland (1996a)
91. Newman, A.J.: Model reduction via the Karhunen-Loève expansion part II: Some elementary examples. Technical Report T.R.96-33, University of Maryland, College Park, Maryland (1996b)
92. Omatu, S., Matumoto, K.: Parameter identification for distributed systems and its application to air pollution estimation. International Journal of Systems Science 22(10), 1993–2000 (1991)
93. Paraskevopoulos, P.N., Bounas, A.C.: Distributed parameter system identification via Walsh-functions. International Journal of Systems Science 9(1), 75–83 (1978)
94. Park, H.M., Cho, D.H.: Low dimensional modeling of flow reactors. International Journal of Heat and Mass Transfer 39(16), 3311–3323 (1996a)
95. Park, H.M., Cho, D.H.: The use of the Karhunen-Loève decomposition for the modeling of distributed parameter systems. Chemical Engineering Science 51(1), 81–98 (1996b)
96. Park, H.M., Kim, T.H., Cho, D.H.: Estimation of parameters in flow reactors using the Karhunen-Loève decomposition. Computers and Chemical Engineering 23(1), 109–123 (1998)
97. Parker, R.S., Heemstra, D., Doyle III, F.J., Pearson, R.K., Ogunnaike, B.A.: The identification of nonlinear models for process control using tailored "plant-friendly" input sequences. Journal of Process Control 11(2), 237–250 (2001)
98. Parlitz, U., Merkwirth, C.: Prediction of spatiotemporal time series based on reconstructed local states. Physical Review Letters 84(9), 1890–1893 (2000)
99. Polis, M.P., Goodson, R.E.: Parameter identification in distributed systems: A synthesizing overview. Proceedings of the IEEE 64(1), 45–61 (1976)
100. Powers, D.L.: Boundary value problems, 4th edn. Academic Press, San Diego (1999)
101. Ranganathan, V., Jha, A.N., Rajamani, V.S.: Identification of linear distributed systems via Laguerre-polynomials. International Journal of Systems Science 15(10), 1101–1106 (1984)
102. Ranganathan, V., Jha, A.N., Rajamani, V.S.: Identification of non-linear distributed systems via a Laguerre-polynomial approach. International Journal of Systems Science 17(2), 241–249 (1986)
103. Rannacher, R., Vexler, B.: A priori error estimates for the finite element discretization of elliptic parameter identification problems with pointwise measurements. SIAM Journal on Control and Optimization 44(5), 1844–1863 (2005)
104. Ray, W.H.: Advanced process control. McGraw-Hill, New York (1981)
105. Romijn, R., Özkan, L., Weiland, S., Ludlage, J., Marquardt, W.: A grey-box modeling approach for the reduction of nonlinear systems. Journal of Process Control 18(9), 906–914 (2008)
106. Rugh, W.: Nonlinear system theory: The Volterral/Wiener approach. Johns Hopkins University Press, Baltimore (1981)
107. Sahan, R.A., Koc-Sahan, N., Albin, D.C., Liakopoulos, A.: Artificial neural network-based modeling and intelligent control of transitional flows. In: Proceeding of the 1997 IEEE International Conference on Control Applications, Hartford, CT, pp. 359–364 (1997)

108. Schetzen, M.: The Volterra and Wiener theories of nonlinear systems. Wiley, New York (1980)
109. Schiesser, W.E.: The numerical method of lines: Integration of partial differential equations. Academic Press, San Diego (1991)
110. Shvartsman, S.Y., Kevrikidis, I.G.: Nonlinear model reduction for control of distributed systems: A computer-assisted study. AIChE Journal 44(7), 1579–1594 (1998)
111. Sirovich, L.: Turbulence and the dynamics of coherent structures parts I-III. Quarterly of Applied Mathematics 45(3), 561–590 (1987)
112. Smaoui, N.: Linear versus nonlinear dimensionality reduction of high-dimensional dynamical systems. SIAM Journal on Scientific Computing 25(6), 2107–2125 (2004)
113. Smaoui, N., Al-Enezi, S.: Modelling the dynamics of nonlinear partial differential equations using neural networks. Journal of Computational and Applied Mathematics 170(1), 27–58 (2004)
114. Stoica, P., Söderström, T.: Instrumental-variable methods for identification of Hammerstein systems. International Journal of Control 35(3), 459–476 (1982)
115. Temam, R.: Infinite-dimensional dynamical systems in mechanics and physics. Springer, New York (1988)
116. Theodoropoulou, A., Adomaitis, R.A., Zafiriou, E.: Model reduction for optimization of rapid thermal chemical vapor deposition systems. IEEE Transactions on Semiconductor Manufacturing 11(1), 85–98 (1998)
117. Uciński, D., Korbicz, J.: Parameter identification of two-dimensional distributed systems. International Journal of Systems Science 21(12), 2441–2456 (1990)
118. Uciński, D., Korbicz, J.: Parameter identification of two-dimensional distributed systems. International Journal of Systems Science 21(12), 2441–2456 (1990)
119. Vande Wouwer, A., Renotte, C., Queinnec, I., Bogaerts, P.H.: Transient analysis of a wastewater treatment biofilter - Distributed parameter modelling and state estimation. Mathematical and Computer Modelling of Dynamical Systems 12(5), 423–440 (2006)
120. Varshney, A., Armaou, A.: Low-order ODE approximations and receding horizon control of surface roughness during thin-film growth. Chemical Engineering Science 63(5), 1246–1260 (2008)
121. Voss, H., Bünner, M.J., Abel, M.: Identification of continuous, spatiotemporal systems. Physical Review E 57(3), 2820–2823 (1998)
122. Webb, A.R.: An approach to non-linear principal components analysis using radially symmetric kernel functions. Journal Statistics and Computing 6(2), 159–168 (1996)
123. Westwick, D., Verhaegen, M.: Identifying MIMO Wiener systems using subspace model identification methods. Signal Processing 52(2), 235–258 (1996)
124. Wilson, D.J.H., Irwin, G.W., Lightbody, G.: RBF principal manifolds for process monitoring. IEEE Transactions on Neural Networks 10(6), 1424–1434 (1999)
125. Zheng, D., Hoo, K.A.: Low-order model identification for implementable control solutions of distributed parameter system. Computers and Chemical Engineering 26(7-8), 1049–1076 (2002)
126. Zheng, D., Hoo, K.A.: System identification and model-based control for distributed parameter systems. Computers and Chemical Engineering 28(8), 1361–1375 (2004)
127. Zheng, D., Hoo, K.A., Piovoso, M.J.: Low-order model identification of distributed parameter systems by a combination of singular value decomposition and the Karhunen-Loève expansion. Industrial & Engineering Chemistry Research 41(6), 1545–1556 (2002)

128. Zhou, X.G., Liu, L.H., Dai, Y.C., Yuan, W.K., Hudson, J.L.: Modeling of a fixed-bed reactor using the KL expansion and neural networks. Chemical Engineering Science 51(10), 2179–2188 (1996)
129. Zhu, Y.C.: Identification of Hammerstein models for control using ASYM. International Journal of Control 73(18), 1692–1702 (2000)
130. Zill, D.G., Cullen, M.R.: Differential equations with boundary-value problems, 5th edn. Brooks/Cole Thomson Learning, Pacific Grove, CA, Australia (2001)

3 Spatio-Temporal Modeling for Wiener Distributed Parameter Systems

Abstract. For Wiener distributed parameter systems (DPS), a spatio-temporal Wiener model (a linear DPS followed by a static nonlinearity) is constructed in this chapter. After the time/space separation, it can be represented by the traditional Wiener system with a set of spatial basis functions. To achieve a low-order model, the Karhunen-Loève (KL) method is used for the time/space separation and dimension reduction. Finally, unknown parameters of the Wiener system are estimated with the least-squares estimation and the instrumental variables method to achieve consistent estimation under noisy measurements. The simulation on the catalytic rod and the experiment on snap curing oven are presented to illustrate the effectiveness of this modeling method.

3.1 Introduction

In the identification of traditional lumped parameter systems (LPS), the block-oriented nonlinear models have been widely used because of their simple structures, abilities to approximate a large class of nonlinear processes and many efficient control schemes. They consist of an interconnection of linear time invariant (LTI) systems and static nonlinearities. Within this class, two common model structures are: the Hammerstein model, which consists of the cascade connection of a static nonlinearity followed by a LTI system, and the Wiener model, in which the order of linear and nonlinear blocks is reversed. This chapter will extend the Wiener system to DPS.

Wiener models are widely used in engineering practice. Modeling the pH neutralization process, the continuous stirred tank reactor and distillation columns are a few examples of their applications. Because a linear structure model can be derived from the block-oriented nonlinear model, the control design and optimization problem of the Wiener model can be easier than that of general nonlinear models (Christofides, 1997; Baker & Christofides, 2000; Armaou & Christofides, 2002; Coca & Billings, 2002; Sahan *et al.*, 1997; Deng, Li & Chen, 2005). The successful examples have been reported in the traditional ordinary differential equation (ODE) processes (Bloemen *et al.*, 2001; Jeong, Yoo & Rhee, 2001; Gerksic *et al.*, 2000; Cervantes, Agamennoni & Figueroa, 2003).

Several approaches to the identification of Wiener models can be found in the literature (Westwick & Verhaegen, 1996; Raich, Zhou & Viberg, 2005; Greblicki, 1994). There are two classes of model parameterization approaches, i.e., nonparametric (Pawlak, Hasiewicz & Wachel, 2007) and parametric approaches (Hagenblad & Ljung, 1998). The parameter identification algorithms can be

H.-X. Li and C. Qi: Spatio-Temporal Modeling of Nonlinear DPS, ISCA 50, pp. 51–72.
springerlink.com

classified into the prediction error method (Wigren, 1993), the subspace method (Gómez, Jutan & Baeyens, 2004), the least-squares (LS) method (Gómez & Baeyens, 2004), the maximum likelihood method (Hagenblad & Ljung, 2000), and the instrumental variables (IV) method (Janczak, 2007), etc. Despite the existence of several well-established Wiener model identification techniques, they are only studied for lumped parameter systems.

In this chapter, a Karhunen-Loève decomposition (KL) based Wiener modeling approach is developed for nonlinear distributed parameter processes. A Wiener distributed parameter system is presented with a distributed linear DPS followed by a static nonlinearity. After the time/space separation, the Wiener distributed parameter system can be represented by a traditional Wiener system with a set of spatial basis functions. To obtain a low-order model, the time/space separation and the dimension reduction are implemented through the Karhunen-Loève decomposition. To identify a Wiener model from temporal coefficients, the least-squares estimation combined with the instrumental variables method is used. This spatio-temporal modeling method can provide consistent parameter estimates under proper conditions. The spatio-temporal Wiener model should have significant approximation capability to nonlinear distributed parameter systems. With the spatio-temporal Wiener model, many control and optimization algorithms (Bloemen *et al.*, 2001; Jeong, Yoo & Rhee, 2001; Gerksic *et al.*, 2000; Cervantes, Agamennoni & Figueroa, 2003; Gómez, Jutan & Baeyens, 2004) developed for the traditional Wiener model can be easily extended to nonlinear distributed parameter processes under unknown circumstances.

This chapter is organized as follows. The Wiener distributed parameter system is given in Section 3.2. The spatio-temporal Wiener modeling problem is described in Section 3.3. In Section 3.4, the Karhunen-Loève decomposition is introduced. The parameterization of the Wiener model is presented in Section 3.5.1. The parameter identification algorithm is given in Section 3.5.2. The numerical simulation and the experiment on snap curing oven are demonstrated in Section 3.6. Finally, a few conclusions are presented in Section 3.7.

3.2 Wiener Distributed Parameter System

A Wiener distributed parameter system is shown in Figure 3.1. The system consists of a distributed linear time-invariant system

$$v(x,t) = G(x,q)u(t) + d(x,t), \tag{3.1}$$

followed by a static nonlinear element $N(\cdot):\mathbb{R}\to\mathbb{R}$, where $G(x,q)$ ($1\times m$) is a transfer function, t is time variable, x is spatial variable defined on the domain Ω, and q stands for the forward shift operator. The input-output relationship of the system is then given by

$$y(x,t) = N(G(x,q)u(t) + d(x,t)), \tag{3.2}$$

where $u(t)\in\mathbb{R}^m$ is the temporal input, $y(x,t)\in\mathbb{R}$ and $d(x,t)\in\mathbb{R}$ are the spatio-temporal output and process noise respectively.

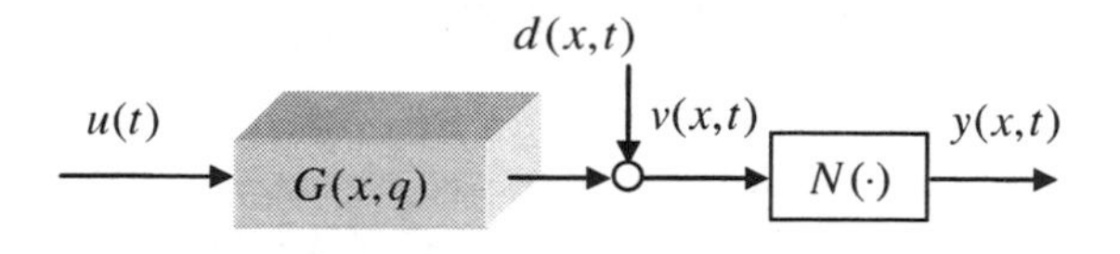

Fig. 3.1 Wiener distributed parameter system

Suppose the transfer function $G(x,q)$ can be expanded onto an infinite number of orthonormal spatial basis functions $\{\varphi_i(x)\}_{i=1}^{\infty}$

$$G(x,q) = \sum_{i=1}^{\infty} \varphi_i(x) G_i(q), \tag{3.3}$$

where $G_i(q)$ ($1 \times m$) is the traditional transfer function. Thus the Wiener distributed parameter system can be represented in Figure 3.2(a) via a time-space separation. Suppose there exists a static nonlinearity $F(\cdot)$ such that

$$N(\varphi(x)v(t)) = \varphi(x)F(v(t)), \tag{3.4}$$

then the Wiener distributed parameter system can be represented by a traditional Wiener system as shown in Figure 3.2(b). This assumption is suitable for a wide range of systems because the nonlinear function F can be different from the function N.

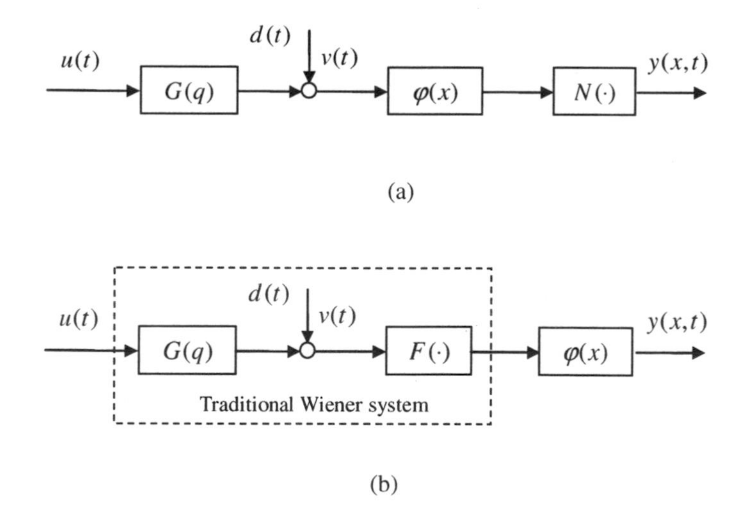

Fig. 3.2 Time/space separation of Wiener distributed parameter system

3.3 Spatio-Temporal Wiener Modeling Methodology

Consider the distributed parameter system in Figure 3.1. Here suppose the system is controlled by the m actuators with implemental temporal signal $u(t)$ and certain spatial distribution. The output is measured at the N spatial locations $x_1,\ldots,x_N$. It should be mentioned that because of the infinite dimensionality of the distributed parameter system, it may require an infinite number of actuators and sensors over the whole space to implement the perfect modeling and control. Due to some practical limitations such as hardware and cost, a limited number of actuators and sensors should be used. The minimal number of actuators and sensors may depend on the process complexity as well as the required accuracy of modeling and control. In this study, we also suppose that the output can be measured without noise, while the system is disturbed by the process noise $d(x,t)$. This assumption is realistic since the influence of (unmeasured) process disturbance is, in general, much greater than that of the measurement noise due to the advances in sensor technologies (Zhu, 2002).

The modeling problem is to identify a proper nonlinear spatio-temporal model from the input $\{u(t)\}_{t=1}^{L}$ and the output $\{y(x_i,t)\}_{i=1,t=1}^{N,L}$, where L is the time length.

As shown in Figure 3.3, the modeling methodology includes two stages. The first stage is the Karhunen-Loève decomposition for the time/space separation. The second stage is the traditional Wiener model identification. Using the time/space synthesis, this model can reconstruct the spatio-temporal dynamics of the system.

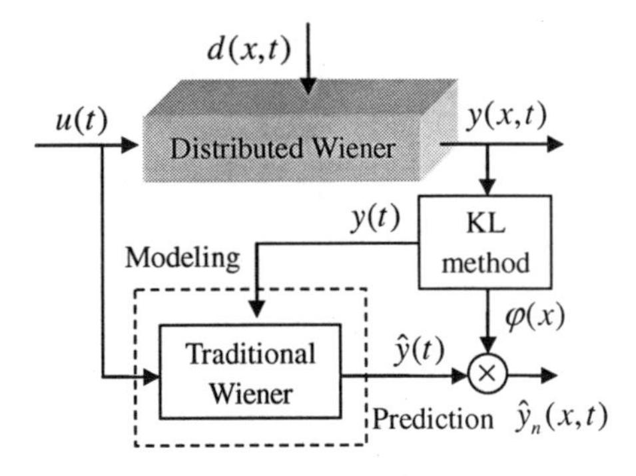

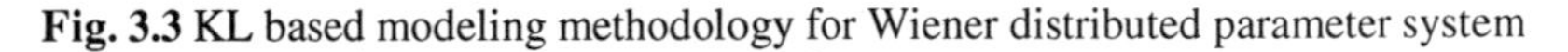
Fig. 3.3 KL based modeling methodology for Wiener distributed parameter system

3.4 Karhunen-Loève Decomposition

Karhunen-Loève expansion (also known as principal component analysis) is to find an optimal basis from a representative set of process data. Suppose we have a set of observations about the process output $\{y(x_i,t)\}_{i=1,t=1}^{N,L}$ (called snapshots) which is assumed to be uniformly sampled in the time and space for simplicity. Here the ensemble average, the norm, and the inner product are defined as

$< f(x,t) >= \frac{1}{L}\sum_{t=1}^{L} f(x,t)$, $\| f(x) \| = (f(x), f(x))^{1/2}$, and $(f(x), g(x)) = \int_{\Omega} f(x)g(x)dx$ respectively.

Motivated by Fourier series, the spatio-temporal variable $y(x,t)$ can be expanded onto an infinite number of orthonormal spatial basis functions $\{\varphi_i(x)\}_{i=1}^{\infty}$

$$y(x,t) = \sum_{i=1}^{\infty} \varphi_i(x) y_i(t) , \tag{3.5}$$

where the temporal coefficients can be computed from

$$y_i(t) = (\varphi_i(x), y(x,t)), i = 1,...,n . \tag{3.6}$$

In practice, it has to be truncated to a finite dimension

$$y_n(x,t) = \sum_{i=1}^{n} \varphi_i(x) y_i(t) , \tag{3.7}$$

where $y_n(x,t)$ denotes the nth-order approximation.

The problem is how to compute the most characteristic structure $\{\varphi_i(x)\}_{i=1}^{n}$ among these snapshots $\{y(x_i,t)\}_{i=1,t=1}^{N,L}$.

Spatial Correlation Method

Actually this problem can be formulated as the one of obtaining a set of functions $\{\varphi_i(x)\}_{i=1}^{n}$ that minimizes the following objective function (Christofides, 2001b):

$$\begin{gathered} \min_{\varphi_i(x)} <\| y(x,t) - y_n(x,t) \|^2> \\ \text{subject to } (\varphi_i, \varphi_i) = 1, \ \varphi_i \in L^2(\Omega), \ i = 1,...,n. \end{gathered} \tag{3.8}$$

The constraint $(\varphi_i, \varphi_i) = 1$ is imposed to ensure that the function $\varphi_i(x)$ is unique. The Lagrangian functional corresponding to this constrained optimization problem is

$$J = <\| y(x,t) - y_n(x,t) \|^2> + \sum_{i=1}^{n} \lambda_i((\varphi_i, \varphi_i) - 1) , \tag{3.9}$$

and the necessary condition of the solution can be obtained as below (Holmes, Lumley & Berkooz, 1996)

$$\int_{\Omega} R(x,\zeta)\varphi_i(\zeta)d\zeta = \lambda_i \varphi_i(x), \ (\varphi_i, \varphi_i) = 1, \ i = 1,...,n , \tag{3.10}$$

where $R(x,\zeta) =< y(x,t)y(\zeta,t) >$ is the spatial two-point correlation function, $\varphi_i(x)$ is the i th eigenfunction, λ_i is the corresponding eigenvalue.

Since the data are always discrete in space, one must solve numerically the integral equation (3.10). Discretizing the integral equation gives a $N \times N$ matrix eigenvalue problem. Thus, at most N eigenfunctions at N sampled spatial locations can be obtained. Then one can use the curve/surface fitting method (Lancaster & Salkauskas, 1986; Dierckx, 1993) interpolate the eigenfunctions to locations where the data are not available.

Temporal Correlation Method

When L is less than N, a computationally efficient way to obtain the solution of (3.10) is provided by the method of snapshots (Sirovich, 1991; Newman, 1996a, 1996b). The eigenfunction $\varphi_i(x)$ is assumed to be expressed as a linear combination of the snapshots as follows

$$\varphi_i(x) = \sum_{t=1}^{L} \gamma_{it} y(x,t) . \tag{3.11}$$

Substituting (3.11) into (3.10) gives the following eigenvalue problem:

$$\int_\Omega \frac{1}{L} \sum_{t=1}^{L} y(x,t) y(\zeta,t) \sum_{k=1}^{L} \gamma_{ik} y(\zeta,k) d\zeta = \lambda_i \sum_{t=1}^{L} \gamma_{it} y(x,t) . \tag{3.12}$$

Define the temporal two-point correlation matrix as C where the element at the tth row and kth column is

$$C_{tk} = \frac{1}{L} \int_\Omega y(\zeta,t) y(\zeta,k) d\zeta . \tag{3.13}$$

Therefore, the $N \times N$ eigenvalue problem (3.10) can be reduced to a $L \times L$ problem as follows

$$C\gamma_i = \lambda_i \gamma_i . \tag{3.14}$$

where $\gamma_i = [\gamma_{i1},...,\gamma_{iL}]^T$ is the ith eigenvector. The solution of the above eigenvalue problem yields the eigenvectors $\gamma_1,...,\gamma_L$, which can be used in (3.11) to construct the eigenfunctions $\varphi_1(x),...,\varphi_L(x)$. Since the matrix C is symmetric and positive semidefinite, its eigenvalues λ_i are real and non-negative. Furthermore, the computed eigenfunctions are orthogonal.

Selection of Dimension n

The maximum number of nonzero eigenvalues is $K \le \min(N,L)$. Let the eigenvalues $\lambda_1 > \lambda_2 > \cdots > \lambda_K$ and the corresponding eigenfunctions $\varphi_1(x)$, $\varphi_2(x)$, $\cdots$, $\varphi_K(x)$ in the order of the magnitude of the eigenvalues. It can be proved that (Holmes, Lumley & Berkooz, 1996)

$$\lambda_i = < (y(x,t), \varphi_i(x))^2 > .$$

The eigenfunction that corresponds to the first eigenvalue is considered to be the most "energetic". The total "energy" is defined as being the sum of the eigenvalues.

To each eigenfunction, assign an "energy" percentage based on the associated eigenvalue:

$$E_i = \lambda_i / \sum_{j=1}^{K} \lambda_j .$$

Usually, the sufficient number of eigenfunctions that capture 99% of the system "energy" is used to determine the value of n. Experiences show that only the first few basis functions expansion can represent the dominant dynamics of many parabolic spatio-temporal systems. It can be shown (Holmes, Lumley & Berkooz, 1996) that for some arbitrary set of basis functions $\{\phi_i(x)\}_{i=1}^{n}$,

$$\sum_{i=1}^{n} < (y(\cdot,t),\varphi_i)^2 > = \sum_{i=1}^{n} \lambda_i^2 \geq \sum_{i=1}^{n} < (y(\cdot,t),\phi_i)^2 > .$$

It means that the Karhunen-Loève expansion is optimal on average in the class of representations by linear combination. That is why Karhunen-Loève expansion can give the lowest dimension n.

3.5 Wiener Model Identification

In order to obtain the temporal coefficients of the spatio-temporal output, define the orthogonal projection operator P_w as

$$y_w(t) = P_w y(\cdot,t) = (\varphi_w, \sum_{i=1}^{\infty} \varphi_i(\cdot) y_i(t)) = \sum_{i=1}^{\infty} (\varphi_w, \varphi_i(\cdot)) y_i(t) , \tag{3.15}$$

where $\varphi_w = [\varphi_1, \varphi_2, \cdots, \varphi_n]^T$. Since the basis functions φ_i, ($i = 1,\ldots,\infty$) are orthonormal, the result of this projection is such that $y_w(t) = [y_1(t),\cdots, y_n(t)]^T$. In practical case y_w can be computed from the pointwise data using spline integration. Similarly, we can define the temporal coefficients of the spatio-temporal noise as $d_w(t) = P_w d(\cdot,t) = [d_1(t),\cdots,d_n(t)]^T$.

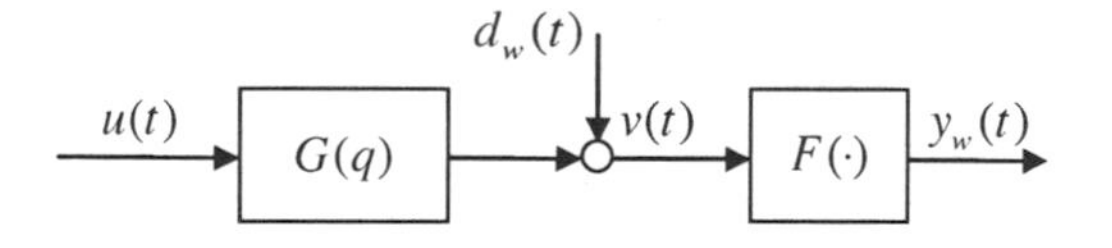

Fig. 3.4 Wiener model

Suppose that the dynamics between $u(t)$ and $y_w(t)$ can be described by a Wiener model. As shown in Figure 3.4, the Wiener model consists of the cascade of a linear time-invariant dynamical system with a $n \times m$ transfer function matrix $G(q)$ (q is the time-shift operator ($qu(t) = u(t + \Delta t)$), followed by a nonlinear static element $F(\cdot):\mathbb{R}^n \to \mathbb{R}^n$. The input-output relationship of the Wiener model is then given by

$$y_w(t) = F(v(t)) = F(G(q)u(t) + d_w(t)), \tag{3.16}$$

where $u(t) \in \mathbb{R}^m$, $y_w(t) \in \mathbb{R}^n$, $v(t) \in \mathbb{R}^n$, and $d_w(t) \in \mathbb{R}^n$, represent the temporal input, output, intermediate variable and process noise at time t, respectively.

Now the identification problem is to estimate the Wiener model from the data set $\{u(t), y_w(t)\}_{t=1}^{L}$.

3.5.1 Model Parameterization

It will be assumed that the transfer function matrix of the linear subsystem is represented as an orthonormal basis expansion of the following form

$$G(q) = \sum_{i=1}^{n_\alpha} \alpha_i g_i(q), \tag{3.17}$$

where $g_i(q)$ ($i = 1,...,n_\alpha$) are known scalar basis functions, $\alpha_i \in R^{n\times m}$ ($i = 1,...,n_\alpha$) are unknown matrix parameters. $g_i(q)$ can be selected as finite impulse response, Laguerre (Wahlberg, 1991) and Kautz functions (Wahlberg, 1994) and generalized orthonormal basis functions (Heuberger, Van den Hof & Bosgra, 1995). Here, Laguerre functions are chosen for the development because of their simplicity, that is they are completely determined by a single parameter. The transfer functions of Laguerre functions are given by

$$g_i(q) = \frac{\sqrt{1-\xi^2}}{q-\xi}\left(\frac{1-\xi q}{q-\xi}\right)^{i-1}, i = 1,2,\cdots,\infty, \tag{3.18}$$

where ξ ($|\xi|<1$) is a stable pole. It can be shown (Wahlberg, 1991) that if this pole is set close to the dominant dynamics of the system to be modeled, a significant reduction in the number of functions and parameters needed to approximate the system with arbitrary accuracy can be achieved.

The intermediate variable $v(t)$ can be written as

$$v(t) = G(q)u(t) + d_w(t) = \sum_{i=1}^{n_\alpha} \alpha_i g_i(q)u(t) + d_w(t). \tag{3.19}$$

On the other hand, the nonlinear function $F(\cdot)$ will be assumed to be invertible, and its inverse $F^{-1}(\cdot)$ can be described as

$$v(t) = F^{-1}(y_w(t)) = \sum_{j=1}^{n_\beta} \beta_j f_j(y_w(t)), \tag{3.20}$$

where $f_j(\cdot):\mathbb{R}^n \to \mathbb{R}^n$ ($j = 1,...,n_\beta$) are nonlinear basis functions such as polynomials, splines, radial basis functions and wavelets (Sjöberg *et al.*, 1995), and $\beta_j \in \mathbb{R}^{n\times n}$

$(j=1,\ldots,n_\beta)$ are unknown matrix parameters. Typically, the polynomial representation is chosen because it is simple to implement and analyze. The assumption of invertible nonlinearities is common in most existing Wiener identification algorithms (Hagenblad, 1999) because it is particularly convenient for control system design, whereas others (Wigren, 1994; Lacy & Bernstein, 2003) allow noninvertible nonlinearities. The orders n_α and n_β are assumed to be known.

Under these assumptions, the identification problem is to estimate the unknown parameter matrices α_i ($i=1,\ldots,n_\alpha$) and β_j $(j=1,\ldots,n_\beta)$ from the data set $\{u(t),y_w(t)\}_{t=1}^L$. Once β_j $(j=1,\ldots,n_\beta)$ are obtained, $F^{-1}(\cdot)$ is known, and thus $F(\cdot)$ can be directly obtained from $F^{-1}(\cdot)$.

3.5.2 Parameter Estimation

In order to obtain the uniqueness of the parameterization, without loss of generality, it will also be assumed that $\beta_1 = I_n$, with I_n standing for a $n\times n$ identity matrix.

Combining (3.19) and (3.20), the following equation is obtained

$$f_1(y_w(t)) = \sum_{i=1}^{n_\alpha}\alpha_i g_i(q)u(t) - \sum_{j=2}^{n_\beta}\beta_j f_j(y_w(t)) + d_w(t)\,, \tag{3.21}$$

which is a linear regression.

Defining

$$\theta = [\alpha_1,\ldots,\alpha_{n_\alpha},\beta_2,\ldots,\beta_{n_\beta}]^T \in \mathbb{R}^{(mn_\alpha+nn_\beta-n)\times n}, \tag{3.22}$$

$$\phi(t) = [(g_1(q)u(t))^T,\ldots,(g_{n_\alpha}(q)u(t))^T,-f_2^T(y_w(t)),\ldots,-f_{n_\beta}^T(y_w(t))]^T \in \mathbb{R}^{mn_\alpha+nn_\beta-n}, \tag{3.23}$$

then (3.21) can be written as

$$f_1(y_w(t)) = \theta^T\phi(t) + d_w(t)\,. \tag{3.24}$$

Now, an estimate $\hat{\theta}$ of θ can be computed by minimizing a quadratic criterion on the prediction errors

$$\hat{\theta} = \arg\min_\theta\{\frac{1}{L}\sum_{t=1}^{L}\| f_1(y_w(t)) - \theta^T\phi(t)\|^2\}\,. \tag{3.25}$$

It is well known that the least-squares estimate is given by

$$\hat{\theta}_{LS} = (\frac{1}{L}\sum_{t=1}^{L}\phi(t)\phi^T(t))^{-1}(\frac{1}{L}\sum_{t=1}^{L}\phi(t)f_1^T(y_w(t)))\,, \tag{3.26}$$

provided that the indicated inverse exists.

The consistency of the estimate $\hat{\theta}_{LS}$ in (3.26), can only be guaranteed in the noise free case, since the regressor $\phi(t)$ at time t will be correlated with the disturbance $d_w(t)$ at the same instant, even if the disturbance is a white noise process.

The instrumental variables method is a simple and effective tool to obtain the consistency estimation (Janczak, 2007; Hagenblad, 1999). The idea of the IV method is to project (3.24) onto the so called instrumental variables vector $\psi(t)$ which is designed properly. Then we have

$$\hat{\theta}_{IV} = (\frac{1}{L}\sum_{t=1}^{L}\psi(t)\phi^T(t))^{-1}(\frac{1}{L}\sum_{t=1}^{L}\psi(t)f_1^T(y_w(t))) . \tag{3.27}$$

Thus the following conditions

$$E[\psi(t)\phi^T(t)] \text{ be nonsingular,} \tag{3.28}$$

$$E[\psi(t)d^T(t)] = 0 , \tag{3.29}$$

are necessary to obtain the consistent parameter estimates. That means the instrumental variables should be chosen such that they are correlated with regression variables $\phi(t)$ but uncorrelated with the disturbance $d_w(t)$. The variance of the IV parameter estimate depends on the choice of instrumental variables. A higher correlation between $\psi(t)$ and $\phi(t)$ results in a smaller variance error. Clearly, the good instrumental variables would contain the undisturbed system outputs but these are not available for measurement. Instead, the outputs of the model, calculated with the least-squares method, can construct the required instrumental variables as below

$$\psi(t) = [(g_1(q)u(t))^T, \cdots, (g_{n_\alpha}(q)u(t))^T, -f_2^T(\hat{y}_w(t)), \cdots, -f_{n_\beta}^T(\hat{y}_w(t))]^T \in \mathbb{R}^{mn_\alpha + nn_\beta - n} . \tag{3.30}$$

Now, estimates of the parameters α_i ($i = 1,...,n_\alpha$) and β_j ($j = 2,...,n_\beta$) can be computed by partitioning the estimate $\hat{\theta}_{IV}$, according to the definition of θ in (3.22).

The identification algorithm can then be summarized as follows.

Algorithm 3.1:

Step 1: Use the measured output $\{y(x_i,t)\}_{i=1,t=1}^{N,L}$ as snapshots, find the spatial basis functions $\{\varphi_i(x)\}_{i=1}^{n}$ via Karhunen-Loève decomposition using (3.10) or (3.14), calculate the temporal coefficients $\{y_w(t)\}_{t=1}^{L}$ using (3.15).

Step 2: Select the proper Laguerre pole ξ and nonlinear basis functions $f_j(\cdot)$, set the system orders n_α and n_β .

Step 3: Compute the least-squares estimate $\hat{\theta}_{LS}$ of θ as in (3.26), then obtain the parameters α_i^{LS} ($i = 1,...,n_\alpha$) and β_j^{LS} ($j = 2,...,n_\beta$) by partitioning the estimate $\hat{\theta}_{LS}$, according to the definition of θ in (3.22).

Step 4: Simulate the model (3.16) with α_i^{LS} ($i = 1,...,n_\alpha$) and β_j^{LS} ($j = 2,...,n_\beta$) to construct the instrumental variables vector $\psi(t)$ in (3.30).

Step 5: Estimate the parameter $\hat{\theta}_{IV}$ with the IV method using (3.27), then compute the parameters α_i^{IV} $(i=1,...,n_\alpha)$ and β_j^{IV} $(j=2,...,n_\beta)$ by partitioning the estimate $\hat{\theta}_{IV}$, according to the definition of θ in (3.22).

Remark 3.1: ***Unknown System Orders***

In the practical situation, the orders n_α and n_β are not known. Also, it may only obtain suboptimal results if the model orders are less than system orders. However, if the upper bounds on the orders are known, then the bounds can be used in (3.17) and (3.20) at the expense of the increasing computational complexity. Alternatively, the performance of F and G can be improved as increasing the system orders until the satisfactory performance is achieved, at the expense of the increasing computational load at each increment.

Remark 3.2: ***Determination of Laguerre Pole***

In order to obtain the significant performance, the parameter ξ is usually obtained from trials. The systematic approach for optimal selection of the Laguerre pole was proposed for the linear system (Fu & Dumont, 1993) and the nonlinear Volterra system (Campello, Favier & Amaral, 2004). For the Wiener system, one approach is the iterative optimization of $\hat{\theta}_{IV}$ and ξ. However, it will turn into a complex nonlinear optimization problem which will not be studied here.

Remark 3.3: ***State-space Realization***

Usually a state-space realization of the identified model is required for control purpose. It can be constructed using existing results on the state-space realizations for orthonormal basis systems (Gómez, 1998).

3.6 Simulation and Experiment

In order to evaluate the presented modeling method, the simulation on a typical distributed process: a catalytic rod is studied first, and then the experiment and modeling for the snap curing oven are presented.

The two models to be compared are stated as follows:

- Spline functions based Wiener (SP-Wiener) model,
- Karhunen-Loève based Wiener (KL-Wiener) model.

The SP-Wiener model is constructed by replacing Karhunen-Loève basis functions φ in (3.15) with spline functions during the modeling procedure. See the reference (Coca & Billings, 2002; Shikin & Plis, 1995) for details on the construction of spline functions.

Define $y(x,t)$ and $\hat{y}_n(x,t)$ as the measured output and the prediction output respectively. Some performance indexes are set up for an easy comparison as follows:

- Spatio-temporal error $e(x,t) = y(x,t) - \hat{y}_n(x,t)$,
- Spatial normalized absolute error $SNAE(t) = \frac{1}{N}\sum_{i=1}^{N} |e(x_i,t)|$,
- Temporal normalized absolute error $TNAE(x) = \sum |e(x,t)| / \sum \Delta t$,
- Root of mean squared error $RMSE = (\frac{1}{NL}\sum_{i=1}^{N}\sum_{t=1}^{L} e(x_i,t)^2)^{1/2}$.

3.6.1 Catalytic Rod

Consider the catalytic rod given in Section 1.1.2. The mathematical model which describes the spatio-temporal evolution of the dimensionless rod temperature consists of the following parabolic PDE (Christofides, 2001b):

$$\frac{\partial y(x,t)}{\partial t} = \frac{\partial^2 y(x,t)}{\partial x^2} + \beta_T (e^{-\frac{\gamma}{1+y}} - e^{-\gamma}) + \beta_u (b(x)^T u(t) - y(x,t)) + d(x,t), \quad (3.31)$$

subject to the boundary and initial conditions:

$$y(0,t) = 0, \ y(\pi,t) = 0, \ y(x,0) = y_0(x),$$

where $y(x,t)$, $u(t)$, $b(x)$, β_T, β_u and γ denote the temperature in the reactor, the manipulated input (temperature of the cooling medium), the actuator distribution, the heat of reaction, the heat transfer coefficient, and the activation energy. $d(x,t)$ is the random process noise. The process parameters are often set as

$$\beta_T = 50, \ \beta_u = 2, \ \gamma = 4.$$

There are available four control actuators $u(t) = [u_1(t), \cdots, u_4(t)]^T$ with the spatial distribution function $b(x) = [b_1(x), \cdots, b_4(x)]^T$, $b_i(x) = H(x-(i-1)\pi/4) - H(x - i\pi/4)$, ($i = 1,...,4$) and $H(\cdot)$ is the standard Heaviside function. There is no unique method to design the input signals for the nonlinear system modeling. In this case, the temporal input $u_i(t) = 1.1 + 5\sin(t/10 + i/10)$ ($i = 1,...,4$) are used for exciting the system. This periodic signal which depends on the spatial location and time instant can excite the nonlinear spatio-temporal dynamics. Unlike the traditional system modeling, due to the infinite-dimensional feature, sufficient sensors should be used to measure the representative spatial features of the distributed parameter system. It depends on the complexity of the spatial dynamics as well as the required modeling accuracy. In general, the number of output measurements is larger than the number of actuators. In this case, twenty sensors uniformly distributed in the space are used for measurements. The random process noise $d(x,t)$ is bounded by 0.01 with zero mean. The PDE system (3.31) is solved using a high-order finite difference method. This solution is used to represent the real system response. See the reference

(Strikwerda, 1989) for details on the finite difference method. The sampling interval Δt is 0.01 and the simulation time is 6. Totally a set of 600 data is collected. The first 500 data is used as the training data with the first 400 data as the estimation data and the next 100 data as the validation data. The validation data is used to monitor the training process and determine some design parameters in the modeling. The remaining 100 data is the testing data.

The process output $y(x,t)$ is shown in Figure 3.5, while the first five Karhunen-Loève basis functions as shown in Figure 3.6 are used for the KL-Wiener model identification. Using the cross-validation method, the temporal bases $g_i(q)$, ($i=1,...,8$) are chosen as Laguerre series with a stable pole $\xi=0.9896$. The nonlinear bases $f_i(y_w)$, ($i=1,...,4$) are designed as the standard polynomials $f_i(y_w)=y_w^i$.

The prediction output $\hat{y}_n(x,t)$ of KL-Wiener model over the whole data set is shown in Figure 3.7, with the prediction error $e(x,t)$ presented in Figure 3.8. It is obvious that KL-Wiener model can reproduce the spatio-temporal dynamics of the original system very well. Now we compare the performance of KL-Wiener model and SP-Wiener model. The SP-Wiener model is established using thirteen third-order splines as shown in Figure 3.9 which are generated using Spline Toolbox of MATLAB. Figure 3.10 displays *SNAE*(t) of these two models over the whole data set, where the solid line corresponds to KL-Wiener model and the dashed line to SP-Wiener model. The *RMSE* of SP-Wiener and KL-Wiener models are about 0.0042402 and 0.0009848 respectively. It is apparent that KL-Wiener model performs much better than SP-Wiener model even if SP-Wiener model allows the use of more number of basis functions. This is owing to the optimal feature of the Karhunen-Loève decomposition. The spatio-temporal KL-Wiener model is very efficient which is suitable for the nonlinear distributed parameter process.

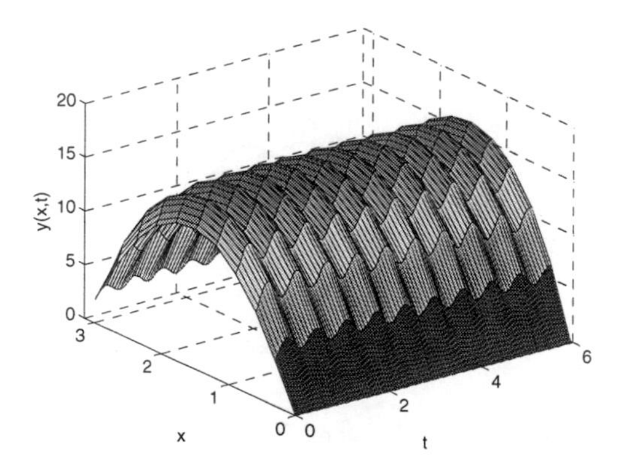

Fig. 3.5 Catalytic rod: Measured output for Wiener modeling

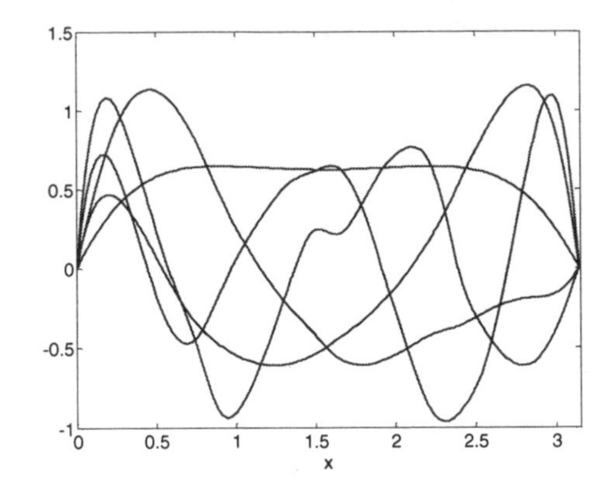

Fig. 3.6 Catalytic rod: KL basis functions for KL-Wiener modeling

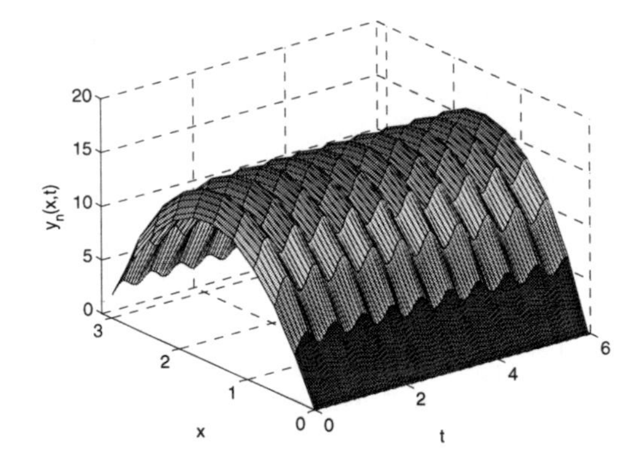

Fig. 3.7 Catalytic rod: KL-Wiener model output

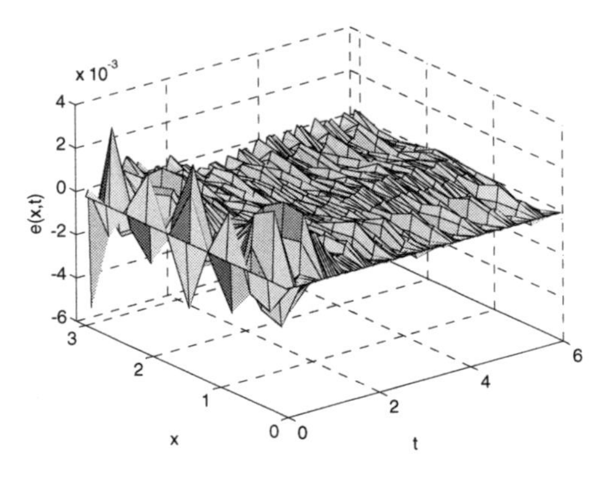

Fig. 3.8 Catalytic rod: Prediction error of KL-Wiener model

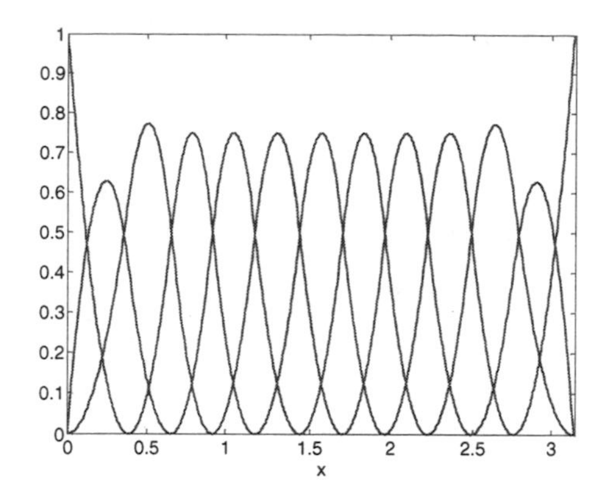

Fig. 3.9 Catalytic rod: Spline basis functions for SP-Wiener modeling

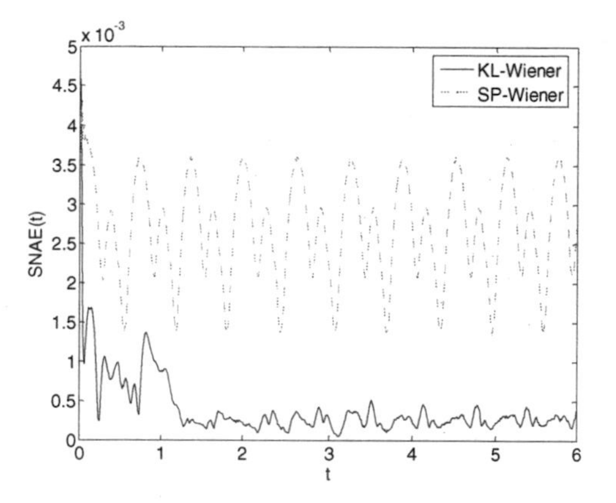

Fig. 3.10 Catalytic rod: $SNAE(t)$ of SP-Wiener and KL-Wiener models

3.6.2 Snap Curing Oven

Consider the snap curing oven in semiconductor back-end packaging industry provided in Section 1.1.1. As shown in Figure 1.1, it can provide the required curing temperature distribution (Deng, Li & Chen, 2005). The problem is to develop a model to estimate the temperature distribution inside the chamber. As shown in Figure 3.11, it is equipped with four heaters (h1-h4) and sixteen temperature sensors (s1-s16) for modeling. Note that more sensors are added for collecting enough spatio-temporal dynamics information about the temperature field. Though the modeling experiment may need more sensors, once the model has been established, a few sensors will be enough for the prediction and control of the temperature distribution.

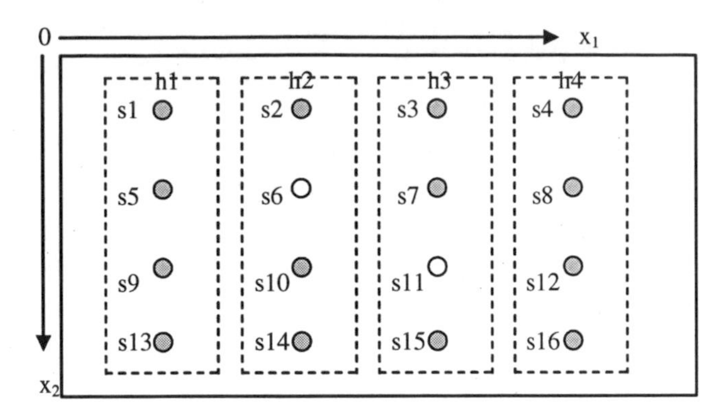

Fig. 3.11 Sensors placement for modeling of snap curing oven

In the experiment, the random input signals are used to excite the thermal process and the first 500 samples for heater 1 are shown in Figure 3.12. A total of 2100 measurements are collected with a sampling interval $\Delta t = 10$ seconds. One thousand and four hundred of measurements from sensors (s1-s5, s7-s10, and s12-s16) are used to estimate the model. The last 700 measurements from sensors (s1-s5, s7-s10, and s12-s16) are chosen to validate the model. All 2100 measurements from the rest sensors (s6, s11), which are not used for training, will be used for testing model performance.

In the spatio-temporal Wiener modeling, five two-dimensional Karhunen-Loève basis functions are used as spatial bases and the first two of them are shown in Figure 3.13 and Figure 3.14. The temporal bases $\phi_i(t)$ are chosen as Laguerre series with the time-scaling factor $p = 0.001$ and the truncation length $q = 3$ using the cross-validation method.

The *KL-Wiener model* is used to model the thermal process. After the training with the first 1400 data from the sensors (s1-s5, s7-s10, and s12-s16), a process model can be obtained with the performance of sensor s1 selected as the example shown in Figure 3.15. Then the model will be tested using the untrained data from sensor s6 and s11. As shown in Figure 3.16, the trained model can perform very well for the untrained data. The predicted temperature distribution of the oven at t=10000s is provided in Figure 3.17.

In order to provide a comparison, a SP-Wiener model is also constructed using twelve third-order splines as spatial basis functions. The first two of them are shown in Figure 3.18 and Figure 3.19. The predicted temperature distribution of the oven at t=10000s is provided in Figure 3.20. The performance index $TNAE(x)$ over the whole data set in Table 3.1 shows that *KL-Wiener model* works much better than *SP-Wiener model.* The *SP-Wiener model* is not suitable for this thermal process because of local spline basis functions used. The effectiveness of the presented modeling method is clearly demonstrated in this real application.

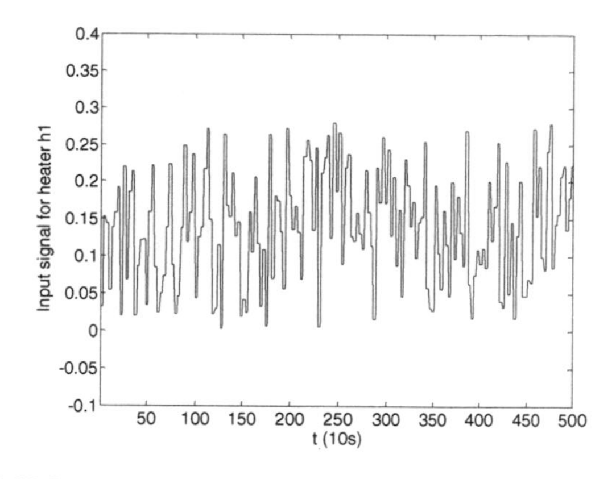

Fig. 3.12 Snap curing oven: Input signals of heater 1 in the experiment

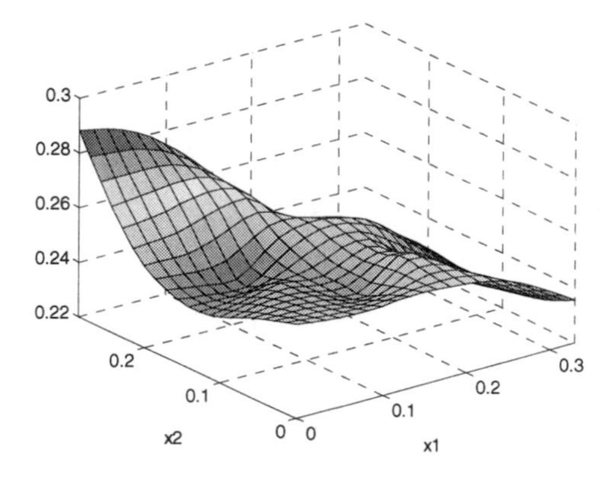

Fig. 3.13 Snap curing oven: KL basis functions (i=1) for KL-Wiener modeling

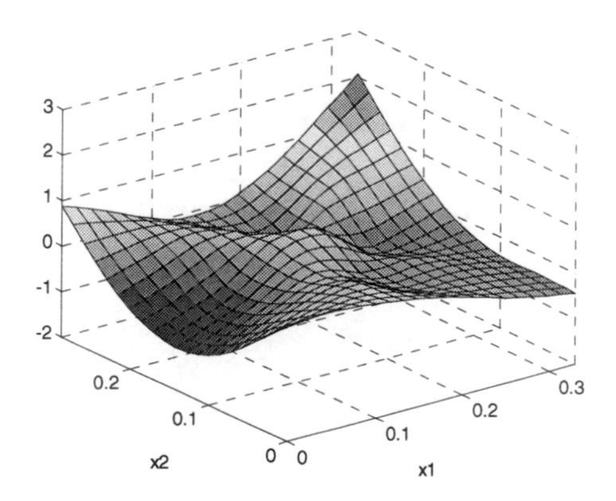

Fig. 3.14 Snap curing oven: KL basis functions (i=2) for KL-Wiener modeling

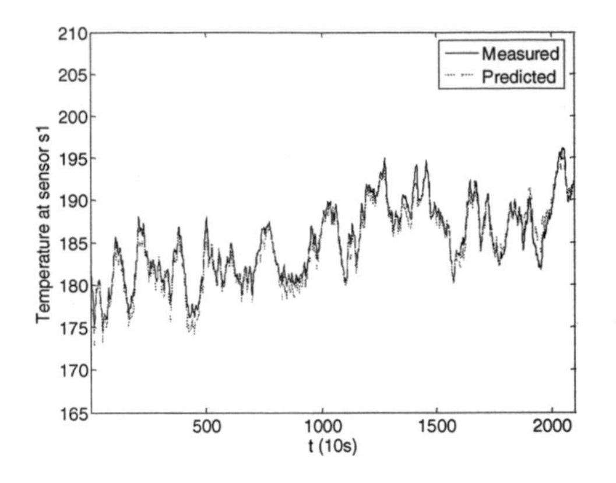

Fig. 3.15 Snap curing oven: Performance of KL-Wiener model at sensor s1

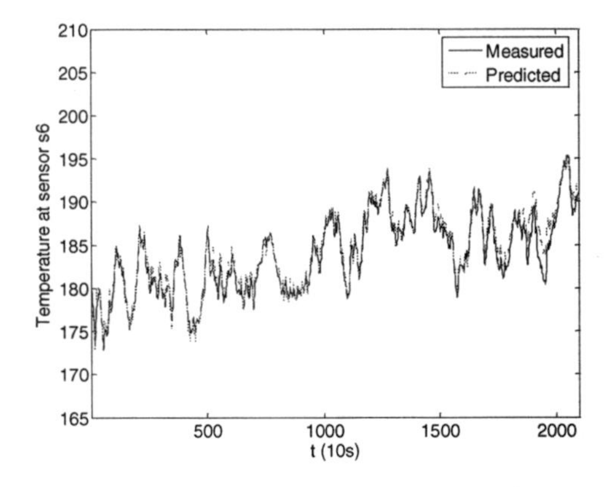

Fig. 3.16 Snap curing oven: Performance of KL-Wiener model at sensor s6

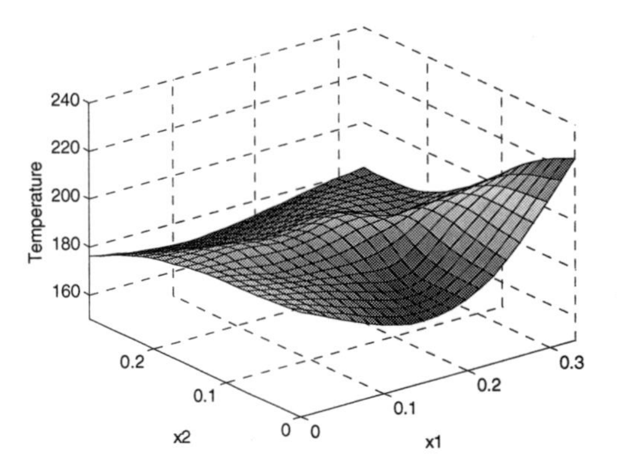

Fig. 3.17 Snap curing oven: Predicted temperature distribution of KL-Wiener model at t=10000s

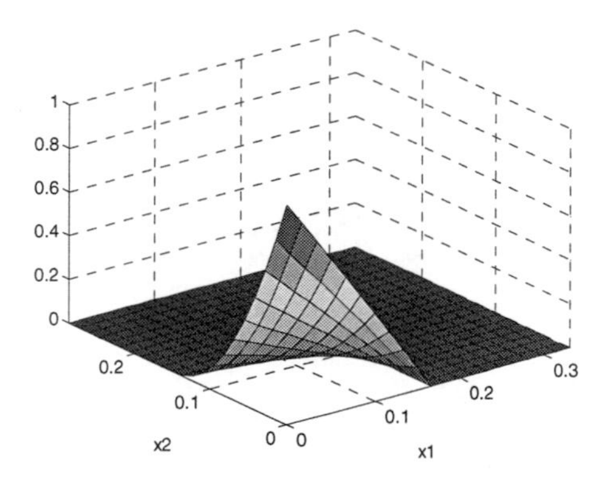

Fig. 3.18 Snap curing oven: Spline basis functions (i=1) for SP-Wiener modeling

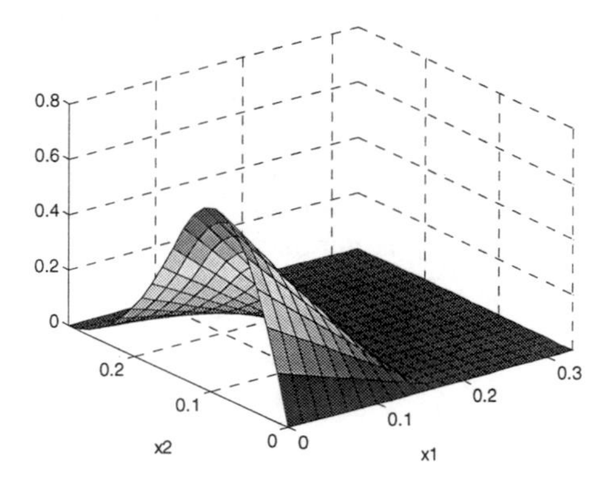

Fig. 3.19 Snap curing oven: Spline basis functions (i=2) for SP-Wiener modeling

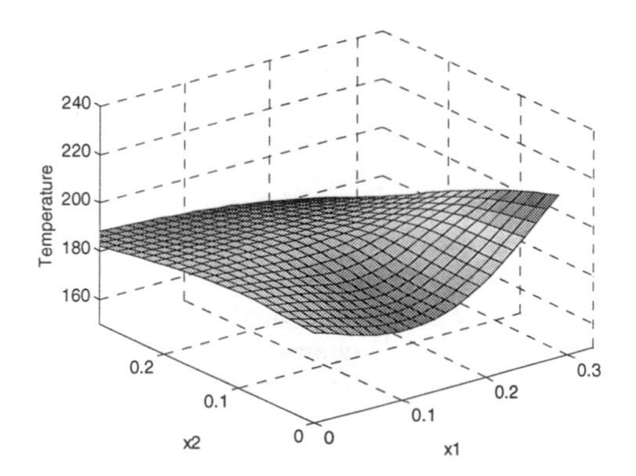

Fig. 3.20 Snap curing oven: Predicted temperature distribution of SP-Wiener model at t=10000s

Table 3.1 Snap curing oven: *TNAE*(x) of KL-Wiener and SP-Wiener models

	s1	s2	s3	s4	s5	s6	s7	s8
KL-Wiener model	0.84	0.88	1.21	0.74	1.81	0.74	0.81	1.21
SP-Wiener model	13.12	14.44	13.01	13.64	12.58	13.82	13.07	15.59
	s9	s10	s11	s12	s13	s14	s15	s16
KL-Wiener model	0.94	0.84	1.05	1.35	0.75	1.04	0.86	0.79
SP-Wiener model	16.39	13.68	13.78	12.82	14.71	13.78	14.51	12.96

3.7 Summary

Traditional Wiener system is extended to DPS. A Wiener distributed parameter system is presented with a distributed linear DPS followed by a static nonlinearity. After the time/space separation, the Wiener distributed parameter system can be represented by traditional Wiener system with a set of spatial basis functions. A KL based spatio-temporal Wiener modeling approach has been presented. The system is assumed to have a finite-dimensional temporal input, as well as a finite-dimensional output measurement at a finite number of spatial locations. The measured output is used to construct a finite-dimensional approximation of the system output which is expanded in terms of Karhunen-Loève spatial basis functions. Using the least-squares estimation and the instrumental variables method, a Wiener model is identified to establish the dynamic relationship between the temporal coefficients and the input. The simulation on the catalytic rod and the experiment on snap curing oven have been presented to illustrate the effectiveness of this modeling method.

References

1. Armaou, A., Christofides, P.D.: Dynamic optimization of dissipative PDE systems using nonlinear order reduction. Chemical Engineering Science 57(24), 5083–5114 (2002)
2. Baker, J., Christofides, P.D.: Finite-dimensional approximation and control of non-linear parabolic PDE systems. International Journal of Control 73(5), 439–456 (2000)
3. Bloemen, H.H.J., Chou, C.T., van den Boom, T.J.J., Verdult, V., Verhaegen, M., Backx, T.C.: Wiener model identification and predictive control for dual composition control of a distillation column. Journal of Process Control 11(6), 601–620 (2001)
4. Campello, R.J.G.B., Favier, G., Amaral, W.C.: Optimal expansions of discrete-time Volterra models using Laguerre functions. Automatica 40(5), 815–822 (2004)
5. Cervantes, A.L., Agamennoni, O.E., Figueroa, J.L.: A nonlinear model predictive control system based on Wiener piecewise linear models. Journal of Process Control 13(7), 655–666 (2003)
6. Christofides, P.D.: Nonlinear and robust control of PDE systems: Methods and applications to transport-reaction processes. Birkhäuser, Boston (2001b)

7. Christofides, P.D., Daoutidis, P.: Finite-dimensional control of parabolic PDE systems using approximate inertial manifolds. Journal of Mathematical Analysis and Applications 216(2), 398–420 (1997)
8. Coca, D., Billings, S.A.: Identification of finite dimensional models of infinite dimensional dynamical systems. Automatica 38(11), 1851–1865 (2002)
9. Deng, H., Li, H.-X., Chen, G.: Spectral-approximation-based intelligent modeling for distributed thermal processes. IEEE Transactions on Control Systems Technology 13(5), 686–700 (2005)
10. Dierckx, P.: Curve and surface fitting with splines. Clarendon, Oxford (1993)
11. Fu, Y., Dumont, G.A.: An optimum time scale for discrete Laguerre network. IEEE Transactions on Automatic Control 38(6), 934–938 (1993)
12. Gerksic, S., Juricic, D., Strmcnik, S., Matko, D.: Wiener model based nonlinear predictive control. International Journal of Systems Science 31(2), 189–202 (2000)
13. Gómez, J.C.: Analysis of dynamic system identification using rational orthonormal bases. PhD Thesis, The University of Newcastle, Australia (1998)
14. Gómez, J.C., Baeyens, E.: Identification of block-oriented nonlinear systems using orthonormal bases. Journal of Process Control 14(6), 685–697 (2004)
15. Gómez, J.C., Jutan, A., Baeyens, E.: Wiener model identification and predictive control of a pH neutralisation process. IEE Proceedings-Control Theory and Applications 151(3), 329–338 (2004)
16. Greblicki, W.: Nonparametric identification of Wiener systems by orthogonal series. IEEE Transactions on Automatic Control 39(10), 2077–2086 (1994)
17. Hagenblad, A.: Aspects of the identification of Wiener models. Thesis No.793, Linköping University, Sweden (1999)
18. Hagenblad, A., Ljung, L.: Maximum likelihood identification of Wiener models with a linear regression initialization. In: Proceedings of the 37th IEEE Conference Decision & Control, Tampa, Fourida, USA, pp. 712–713 (1998)
19. Hagenblad, A., Ljung, L.: Maximum likelihood estimation of Wiener models. Report no.: LiTH-ISY-R-2308, Linköping University, Sweden (2000)
20. Heuberger, P.S.C., Van den Hof, P.M.J., Bosgra, O.H.: A generalized orthonormal basis for linear dynamical systems. IEEE Transactions on Automatic Control 40(3), 451–465 (1995)
21. Holmes, P., Lumley, J.L., Berkooz, G.: Turbulence, coherent structures, dynamical systems, and symmetry. Cambridge University Press, New York (1996)
22. Janczak, A.: Instrumental variables approach to identification of a class of MIMO Wiener systems. Nonlinear Dynamics 48(3), 275–284 (2007)
23. Jeong, B.-G., Yoo, K.-Y., Rhee, H.-K.: Nonlinear model predictive control using a Wiener model of a continuous methyl methacrylate polymerization reactor. Industrial and Engineering Chemistry Research 40(25), 5968–5977 (2001)
24. Lacy, S.L., Bernstein, D.S.: Identification of FIR Wiener systems with unknown, non-invertible, polynomial non-linearities. International Journal of Control 76(15), 1500–1507 (2003)
25. Lancaster, P., Salkauskas, K.: Curve and surface fitting: An introduction. Academic Press, London (1986)
26. Newman, A.J.: Model reduction via the Karhunen-Loève expansion part I: An exposition. Technical Report T.R.96-32, University of Maryland, College Park, Maryland (1996a)

27. Newman, A.J.: Model reduction via the Karhunen-Loève expansion part II: Some elementary examples. Technical Report T.R.96-33, University of Maryland, College Park, Maryland (1996b)
28. Pawlak, M., Hasiewicz, Z., Wachel, P.: On nonparametric identification of Wiener systems. IEEE Transactions on Signal Processing 55(2), 482–492 (2007)
29. Raich, R., Zhou, G.T., Viberg, M.: Subspace based approaches for Wiener system identification. IEEE Transactions on Automatic Control 50(10), 1629–1634 (2005)
30. Sahan, R.A., Koc-Sahan, N., Albin, D.C., Liakopoulos, A.: Artificial neural network-based modeling and intelligent control of transitional flows. In: Proceeding of the 1997 IEEE International Conference on Control Applications, Hartford, CT, pp. 359–364 (1997)
31. Shikin, E.V., Plis, A.I.: Handbook on splines for the user. CRC Press, Boca Raton (1995)
32. Sirovich, L.: New perspectives in turbulence, 1st edn. Springer, New York (1991)
33. Sjöberg, J., Zhang, Q., Ljung, L., Benveniste, A., Delyon, B., Glorennec, P., Hjalmarsson, H., Juditsky, A.: Nonlinear black-box modeling in system identification: A unified approach. Automatica 31(12), 1691–1724 (1995)
34. Strikwerda, J.C.: Finite difference schemes and partial differential equations. Wads. & Brooks/Cole Adv. Bks. & S.W., Pacific Grove (1989)
35. Wahlberg, B.: System identification using Laguerre models. IEEE Transactions on Automatic Control 36(5), 551–562 (1991)
36. Wahlberg, B.: System identification using Kautz models. IEEE Transactions on Automatic Control 39(6), 1276–1282 (1994)
37. Westwick, D., Verhaegen, M.: Identifying MIMO Wiener systems using subspace model identification methods. Signal Processing 52(2), 235–258 (1996)
38. Wigren, T.: Recursive prediction error identification using the nonlinear Wiener model. Automatica 29(4), 1011–1025 (1993)
39. Wigren, T.: Convergence analysis of recursive identification algorithms based on the nonlinear Wiener model. IEEE Transactions on Automatic Control 39(11), 2191–2206 (1994)
40. Zhu, Y.C.: Estimation of an N-L-N Hammerstein-Wiener model. Automatica 38(9), 1607–1614 (2002)

4 Spatio-Temporal Modeling for Hammerstein Distributed Parameter Systems

Abstract. A spatio-temporal Hammerstein modeling approach is presented in this chapter. To model the nonlinear distributed parameter system (DPS), a spatio-temporal Hammerstein model (a static nonlinearity followed by a linear DPS) is constructed. After the time/space separation, it can be represented by the traditional Hammerstein system with a set of spatial basis functions. To achieve a low-order model, the Karhunen-Loève (KL) method is used for the time/space separation and dimension reduction. Then a compact Hammerstein model structure is determined by the orthogonal forward regression, and their unknown parameters are estimated with the least-squares method and the singular value decomposition. The simulation and experiment are presented to show the effectiveness of this spatio-temporal modeling method.

4.1 Introduction

Hammerstein models are widely used in engineering practice due to their capability of approximating many nonlinear industrial processes and simple block-oriented nonlinear structure (i.e., a nonlinear static block in series with a linear dynamic system). Examples include modeling of the pH neutralization process, the continuous stirred tank reactor and distillation columns. Because a linear structure model can be derived from the block-oriented nonlinear structure, the linear control design can be easily extended to Hammerstein models. Successful control applications have been reported for traditional ordinary differential equation (ODE) processes (Fruzzetti, Palazoglu & McDonald, 1997; Samuelsson, Norlander & Carlsson, 2005). However, because the traditional Hammerstein model does not have inherent capability to process spatio-temporal information, few studies have been found in its application in the distributed parameter system (DPS).

Many approaches can be found in the identification of Hammerstein models (e.g., Narendra & Gallman, 1966; Stoica & Söderström, 1982; Bai, 1998; Chen, 2004; Zhu, 2000; Greblicki, 2006; Vörös, 2003; Gómez & Baeyens, 2004). It is notable that an algorithm based on the least-squares estimation and the singular value decomposition (LSE-SVD) is proposed for Hammerstein-Wiener systems (Bai, 1998) and extensively studied for Hammerstein systems (Gómez & Baeyens, 2004). This algorithm can avoid the local minima since it does not require any nonlinear optimization. However, the model structure is assumed to be known in advance and only unknown parameters need to be estimated (Bai 1998). In many cases, the structure is often unknown and the model terms and orders have to be determined carefully. If the structure is inappropriate, it is very difficult to

H.-X. Li and C. Qi: Spatio-Temporal Modeling of Nonlinear DPS, ISCA 50, pp. 73–94.
springerlink.com

guarantee the modeling performance. In Gómez & Baeyens (2004), the order of linear part is determined using the linear subspace identification method and the order of nonlinear part is based on the cross-validation technique. This separated order selection may not provide a compact Hammerstein model since the term selection problem is not considered.

Because the number of possible model terms may be very large, it may lead to a very complex model and an ill-condition problem. In this chapter, we want to obtain a parsimonious model for control which should be as simple as possible. In fact, many terms are redundant and only a small number of important terms are necessary to describe the system with a given accuracy. The term selection problem has been extensively studied for the linear regression model (e.g., Haber & Unbehauen, 1990; Piroddi & Spinelli, 2003; Lind & Ljung, 2008; Billings *et al.*, 1988a, 1988b). In particular, the orthogonal forward regression (OFR) (Billings *et al.*, 1988a, 1988b) is a fast and effective algorithm to determine significant model terms among a candidate set. Here we will extend the OFR to the Hammerstein model identification.

In this chapter, a Karhunen-Loève (KL) decomposition based Hammerstein modeling approach is developed for unknown nonlinear distributed parameter processes with the spatio-temporal output. A Hammerstein distributed parameter system is presented with a static nonlinearity followed by a linear DPS. After the time/space separation, this Hammerstein distributed parameter system can be represented by the traditional Hammerstein system with a set of spatial basis functions. Firstly, the KL decomposition is used for the time/space separation, where a few dominant spatial basis functions are estimated from the spatio-temporal data and the low-dimensional temporal coefficients are obtained simultaneously. Secondly, a low-order and parsimonious Hammerstein model is identified from the low-dimensional temporal data to establish the system dynamics, where the compact or sparse model structure is determined by the orthogonal forward regression algorithm, and the parameters are estimated using the least-squares method and the singular value decomposition. The presented time/space separated Hammerstein model has significant approximation capability to many nonlinear distributed parameter systems. With this model, many control and optimization algorithms designed for traditional Hammerstein model can be extended to nonlinear distributed parameter processes.

This chapter is organized as follows. The Hammerstein distributed parameter system is given in Section 4.2. The spatio-temporal Hammerstein modeling problem and methodology are described in Section 4.3. In Section 4.4, the Karhunen-Loève decomposition based time/space separation is introduced. The parameterization of the Hammerstein model is presented in Section 4.5.1. The model structure selection and the parameter identification algorithm are given in Sections 4.5.2 and 4.5.3. Section 4.6 contains the simulation example and the experiment on the snap curing oven. Finally, a few conclusions are presented in Section 4.7.

4.2 Hammerstein Distributed Parameter System

A Hammerstein distributed parameter system is shown in Figure 4.1. The system consists of a static nonlinear element $N(\cdot):\mathbb{R}^m \to \mathbb{R}^m$ followed by a distributed linear time-invariant system

$$y(x,t) = G(x,q)v(t), \tag{4.1}$$

with a transfer function $G(x,q)$ ($1\times m$), where t is time variable, x is spatial variable defined on the domain Ω, and q stands for the forward shift operator. The input-output relationship of the system is then given by

$$y(x,t) = G(x,q)N(u(t)), \tag{4.2}$$

where $u(t)\in\mathbb{R}^m$ is the temporal input and $y(x,t)\in\mathbb{R}$ is the spatio-temporal output. In this chapter only the single-output (SO) system is considered. The extension of the results to the multi-output (MO) system is straightforward.

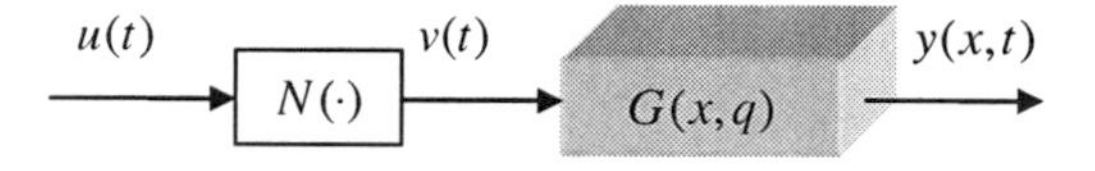

Fig. 4.1 Hammerstein distributed parameter system

Suppose the transfer function $G(x,q)$ can be expanded onto an infinite number of orthonormal spatial basis functions $\{\varphi_i(x)\}_{i=1}^{\infty}$

$$G(x,q) = \sum_{i=1}^{\infty} \varphi_i(x) G_i(q), \tag{4.3}$$

where $G_i(q)$ ($1\times m$) is the traditional transfer function. Thus actually the Hammerstein distributed parameter system can be represented by the traditional Hammerstein system via time-space separation

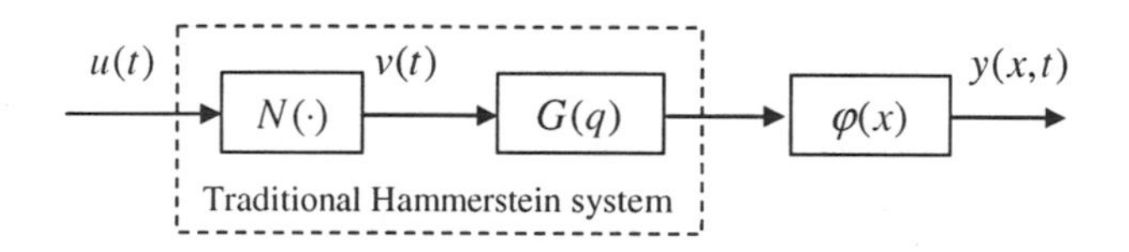

Fig. 4.2 Time/space separation of Hammerstein distributed parameter system

4.3 Spatio-Temporal Hammerstein Modeling Methodology

Consider the nonlinear Hammerstein distributed parameter system in Figure 4.1. Suppose the system is controlled by the m actuators with implemental temporal signal $u(t)$ and certain spatial distribution, and the output is measured at the N spatial locations $x_1, \ldots, x_N$. Because of the infinite dimensionality of the distributed parameter system, it may require an infinite number of actuators and sensors over the whole space to have a perfect modeling and control. Due to practical limitations, a limited number of actuators and sensors have to be used. The number of actuators and sensors may depend on the process complexity, the desired accuracy of modeling and control, physical constraints and cost consideration etc. The modeling problem is to identify a low-order, simple nonlinear and parsimonious spatio-temporal model from the input $\{u(t)\}_{t=1}^{L}$ and the output $\{y(x_i,t)\}_{i=1,t=1}^{N,L}$, where L is the time length.

As shown in Figure 4.3, the modeling methodology includes two stages. The first stage is the Karhunen-Loève decomposition for the time/space separation. The second stage is the traditional Hammerstein model (including the structure and parameters) identification. Using the time/space synthesis, this model can reconstruct the spatio-temporal dynamics of the system.

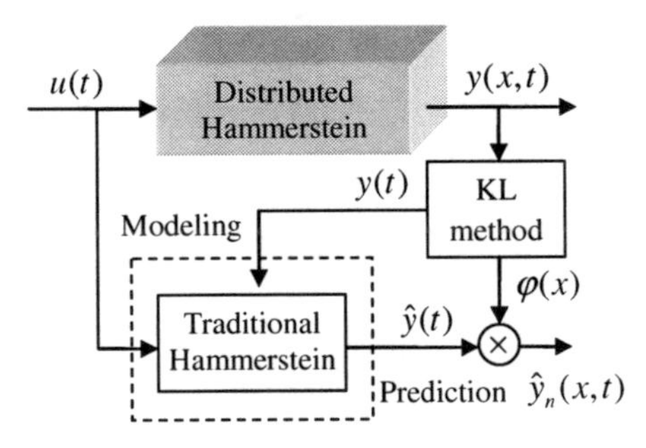

Fig. 4.3 KL based modeling methodology for Hammerstein distributed parameter system

4.4 Karhunen-Loève Decomposition

For simplicity, assume the process output $\{y(x_i,t)\}_{i=1,t=1}^{N,L}$ (called snapshots) is uniformly sampled in the time and space. Here the ensemble average, the inner product and the norm are defined as $\langle f(x,t)\rangle = \frac{1}{L}\sum_{t=1}^{L} f(x,t)$, $(f(x),g(x)) = \int_{\Omega} f(x)g(x)dx$, and $\| f(x)\| = (f(x),f(x))^{1/2}$ respectively.

Motivated by Fourier series, the spatio-temporal variable $y(x,t)$ can be expanded onto an infinite number of orthonormal spatial basis functions $\{\varphi_i(x)\}_{i=1}^{\infty}$

$$y(x,t) = \sum_{i=1}^{\infty} \varphi_i(x) y_i(t), \tag{4.4}$$

Because the spatial basis functions are orthonormal, i.e., $(\varphi_i(x), \varphi_j(x)) = \int_\Omega \varphi_i(x)\varphi_j(x)dx = 0 \ (i \neq j), \ or \ 1 \ (i = j)$, the temporal coefficients can be computed from

$$y_i(t) = (\varphi_i(x), y(x,t)), i = 1, ..., n. \tag{4.5}$$

In practice, it has to be truncated to a finite dimension

$$y_n(x,t) = \sum_{i=1}^{n} \varphi_i(x) y_i(t), \tag{4.6}$$

where $y_n(x,t)$ denotes the nth-order approximation. Similarly a finite order truncation of $G(x,q)$ is given by

$$G_n(x,q) = \sum_{i=1}^{n} \varphi_i(x) G_i(q). \tag{4.7}$$

The main problem of using Karhunen-Loève decomposition for the time/space separation is to compute the most characteristic spatial structure $\{\varphi_i(x)\}_{i=1}^n$ among the spatio-temporal output $\{y(x_i,t)\}_{i=1,t=1}^{N,L}$. The typical $\{\varphi_i(x)\}_{i=1}^n$ can be found by minimizing the following objective function

$$\begin{gathered} \min_{\varphi_i(x)} \ \langle \| y(x,t) - y_n(x,t) \|^2 \rangle \\ \text{subject to } (\varphi_i, \varphi_i) = 1, \ \varphi_i \in L^2(\Omega), \ i = 1, ..., n. \end{gathered} \tag{4.8}$$

The orthonormal constraint $(\varphi_i, \varphi_i) = 1$ is imposed to ensure that the function $\varphi_i(x)$ is unique.

Actually, Karhunen-Loève decomposition can be implemented in several ways, e.g., spatial correlation method and temporal correlation method. Because the Karhunen-Loève expansion is optimal on average in the class of representations by linear combination, it can give the lowest dimension n among all linear expansions. See Section 3.4 for more details about the implementations and the determination of dimension n.

4.5 Hammerstein Model Identification

After identifying the dominant spatial basis functions $\{\varphi_i(x)\}_{i=1}^n$, the corresponding temporal coefficients $\{y_i(t)\}_{i=1}^n$ of the spatial-temporal output $y(x,t)$ can be obtained using (4.5). In practice $\{y_i(t)\}_{i=1}^n$ can be computed from the pointwise data using the spline integration. Define $y(t) = [y_1(t), \cdots, y_n(t)]^T$.

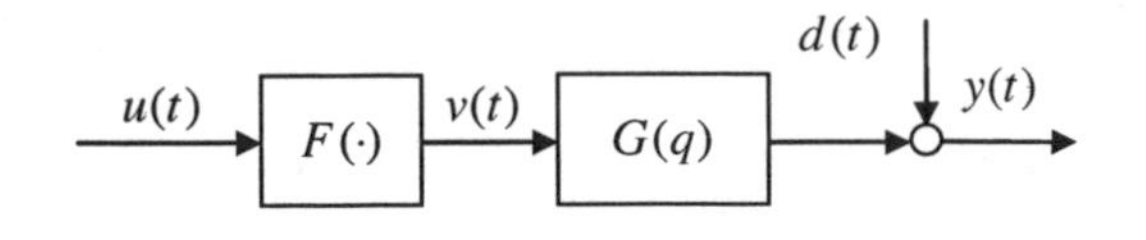

Fig. 4.4 Hammerstein model

Suppose that the dynamics between $u(t)$ and $y(t)$ can be described by a Hammerstein model. As shown in Figure 4.4, the Hammerstein model consists of the cascade of a nonlinear static element $F(\cdot):\mathbb{R}^m \to \mathbb{R}^m$ followed by a linear time-invariant (LTI) dynamical system $G(q)$ (a $n\times m$ transfer function matrix), where q is the time-shift operator ($qu(t)=u(t+\Delta t)$. The input-output relationship of the Hammerstein model is then given by

$$y(t)=G(q)v(t)+d(t)=G(q)F(u(t))+d(t)\,, \tag{4.9}$$

where $u(t)\in\mathbb{R}^m$, $y(t)\in\mathbb{R}^n$, $v(t)\in\mathbb{R}^n$ and $d(t)\in\mathbb{R}^n$ represent the temporal input, output, intermediate variable and modeling error at time t, respectively.

Now the identification is to estimate the Hammerstein model from the low-dimensional temporal data set $\{u(t),y(t)\}_{t=1}^{L}$.

4.5.1 Model Parameterization

Assume that the LTI system $G(q)$ has the ARX form

$$y(t)=A(q^{-1})y(t)+B(q^{-1})v(t)\,, \tag{4.10}$$

where $A(q^{-1})$ and $B(q^{-1})$ are $n\times n$ and $n\times m$ matrix polynomials

$$A(q^{-1})=A_1q^{-1}+\cdots+A_{n_y}q^{-n_y}\,, \tag{4.11}$$

$$B(q^{-1})=B_1q^{-1}+\cdots+B_{n_u}q^{-n_u}\,. \tag{4.12}$$

Here $A_i\in\mathbb{R}^{n\times n}$ $(i=1,\ldots,n_y)$ and $B_i\in\mathbb{R}^{n\times m}$ $(i=1,\ldots,n_u)$ are unknown matrix parameters, n_u and n_y are the maximum input and output lags respectively. On the other hand, assume that the nonlinear static element $F(u)$ can be approximated by

$$v(t)=F(u)=\sum_{i=1}^{n_f}C_if_i(u)\,, \tag{4.13}$$

where $f_i(\cdot):\mathbb{R}^m\to\mathbb{R}^m$ $(i=1,\ldots,n_f)$ are nonlinear basis functions such as polynomials, splines, radial basis functions and wavelets (Sjöberg *et al.*, 1995), $C_i\in\mathbb{R}^{m\times m}$

$(i = 1,\ldots,n_f)$ are unknown matrix parameters, and n_f is the number of basis functions.

Many algorithms have been proposed to identify this class of models when the values of n_y, n_u and n_f are known. However, their true values are often unknown at the beginning of the identification. Moreover, experiences show that often many terms in (4.11)-(4.13) are redundant and can be removed from the model. Therefore, there exist the values of n_{ys}, n_{us} and n_{fs} (generally $n_{ys} \ll n_y$, $n_{us} \ll n_u$ and $n_{fs} \ll n_f$), such that the model

$$y(t) = A_s(q^{-1})y(t) + B_s(q^{-1})v(t), \tag{4.14}$$

$$A_s(q^{-1}) = A_{i_1}q^{-i_1} + \cdots + A_{i_{n_{ys}}}q^{-i_{n_{ys}}}, \tag{4.15}$$

$$B_s(q^{-1}) = B_{j_1}q^{-j_1} + \cdots + B_{j_{n_{us}}}q^{-j_{n_{us}}}, \tag{4.16}$$

$$v = F_s(u) = C_{k_1}f_{k_1}(u) + \cdots + C_{k_{n_{fs}}}f_{k_{n_{fs}}}(u), \tag{4.17}$$

can provide a satisfactory representation over the range considered for the measured input-output data, where $i_r \in \{1,\ldots,n_y\}$, $j_w \in \{1,\ldots,n_u\}$ and $k_v \in \{1,\ldots,n_f\}$. Here we extend the OFR algorithm to determine the compact structure of the Hammerstein model.

4.5.2 Structure Selection

Substituting (4.11)-(4.13) into (4.10), the input-output relationship can be written as

$$y(t) = \sum_{i=1}^{n_y} A_i y(t-i) + \sum_{j=1}^{n_u} B_j \sum_{k=1}^{n_f} C_k f_k(u(t-j)) + e(t), \tag{4.18}$$

where $e(t)$ is the equation error. Define $D_{jk} = B_j C_k \in \mathbb{R}^{n \times m}$, then (4.18) can be rewritten as

$$y(t) = \sum_{i=1}^{n_y} A_i y(t-i) + \sum_{j=1}^{n_u} \sum_{k=1}^{n_f} D_{jk} f_k(u(t-j)) + e(t). \tag{4.19}$$

For every row p in (4.19), we have the following single-output (SO) form

$$y_p(t) = \sum_{i=1}^{n_y} A_i(p,:)y(t-i) + \sum_{j=1}^{n_u} \sum_{k=1}^{n_f} D_{jk}(p,:) f_k(u(t-j)) + e_p(t), \tag{4.20}$$

and

$$y_p(t) = \sum_{i=1}^{n_y}\sum_{w=1}^{n} A_i(p,w)y_w(t-i) + \sum_{j=1}^{n_u}\sum_{k=1}^{n_f}\sum_{l=1}^{m} D_{jk}(p,l)f_{kl}(u(t-j)) + e_p(t), \tag{4.21}$$

where $A(i,:)$ symbolizes the ith row of A and $A(i,j)$ denotes the element at the ith row and jth column of A. (4.21) can be written as a linear regression form

$$y_p(t) = \sum_{i=1}^{M} \phi_i(t)\theta_{pi} + e_p(t), \tag{4.22}$$

where the regressors $\phi_i(t)$ ($i=1,...,M$) are formed from $y_w(t-i)$ and $f_{kl}(u(t-j))$, θ_{pi} ($p=1,...,n$, $i=1,...,M$) are the corresponding parameters $A_i(p,w)$ and $D_{jk}(p,l)$, and $M = nn_y + mn_f n_u$.

Orthogonal Forward Regression

In the OFR algorithm (Billings *et al.*, 1988a, 1988b), all the terms in (4.22) are orthogonalized as below

$$y_p(t) = \sum_{i=1}^{M} \psi_i(t) z_{pi} + e_p(t), \tag{4.23}$$

where $\psi_i(t)$ and z_{pi} denote orthogonal regressors and unknown parameters respectively. Since the regressors are orthogonal, the unknown parameter and error reduction ratio (ERR) can be computed one by one

$$\hat{z}_{pi} = \frac{\sum_{t=1}^{L} \psi_i(t) y_p(t)}{\sum_{t=1}^{L} \psi_i^2(t)},$$

$$err_{pi} = \frac{\sum_{t=1}^{L} [\psi_i(t)\hat{z}_{pi}]^2}{\sum_{t=1}^{L} y_p^2(t)}.$$

The ERR values can give a measure of the significance of each candidate model term.

- *Selection of the first term*

All the terms $\phi_i(t)$, $i=1,\ldots,M$ in (4.22) are considered as the possible candidates for the first significant term in (4.23)

$$\psi_1^{(i)}(t) = \phi_i(t), 1 \le i \le M .$$

Then the parameters and the corresponding error reduction ratios are computed

$$\hat{z}_1^{(pi)} = \frac{\sum_{t=1}^{L} \psi_1^{(i)}(t) y_p(t)}{\sum_{t=1}^{L} [\psi_1^{(i)}(t)]^2}, 1 \le i \le M,$$

$$err_1^{(pi)} = \frac{\sum_{t=1}^{L} [\psi_1^{(i)}(t) \hat{z}_1^{(pi)}]^2}{\sum_{t=1}^{L} y_p^2(t)}, 1 \le i \le M.$$

The term corresponding to the maximum error reduction ratio (e.g., $\psi_j(t)$), is selected as the first significant term $\psi_1(t)$ in (4.23).

- *Selection of the other terms*

All other terms, except $\psi_j(t)$, are considered as candidates to be orthogonalized into (4.23). Compute

$$\psi_2^{(i)}(t) = \phi_i(t) - \alpha_{12}^{(i)} \psi_1(t), 1 \le i \le M, i \ne j,$$

where

$$\alpha_{12}^{(i)} = \frac{\sum_{t=1}^{L} \psi_1(t) \phi_i(t)}{\sum_{t=1}^{L} \psi_1^2(t)}.$$

Then estimate the parameters and compute the corresponding error reduction ratios

$$\hat{z}_2^{(pi)} = \frac{\sum_{t=1}^{L} \psi_2^{(i)}(t) y_p(t)}{\sum_{t=1}^{L} [\psi_2^{(i)}(t)]^2}, 1 \le i \le M, i \ne j,$$

$$err_2^{(pi)} = \frac{\sum_{t=1}^{L} [\psi_2^{(i)}(t) \hat{z}_2^{(pi)}]^2}{\sum_{t=1}^{L} y_p^2(t)}, 1 \le i \le M, i \ne j.$$

The term with the maximum error reduction ratio is then selected as the second term $\psi_2(t)$. Continue the above procedure, and stop at the M_{ps} step until

$$1 - \sum_{i=1}^{M_{ps}} err_{pi} < \rho_e,$$

where ρ_e is a desired error. Because only the most significant term is selected, the OFR algorithm will provide a parsimonious model.

- *Structure design of the Hammerstein model*

From the selected orthogonal model

$$y_p(t) = \sum_{i=1}^{M_{ps}} \psi_i(t) z_{pi} + e_p(t) \text{ ,}$$

it is straightforward to obtain the index set of the corresponding terms $\phi_i(t)$ in the SO model (4.22) as $I_{ps} = \{i_1, ..., i_{M_{ps}}\}$, p=1,..., n. Now define a combination of the selected term index as

$$I_c = I_{1s} \cup I_{2s} \cup \cdots \cup I_{ns} \text{ .}$$

For the multi-output (MO) Hammerstein model (4.19), the index sets of significant terms $I_{ys} = \{i_1, ..., i_{n_{ys}}\}$, $I_{us} = \{j_1, ..., j_{n_{us}}\}$, and $I_{fs} = \{k_1, ..., k_{n_{fs}}\}$ will be determined by I_c. The selection rule is that for each $i \in \{1, ..., n_y\}$, $j \in \{1, ..., n_u\}$ and $k \in \{1, ..., n_f\}$, if any element of the corresponding model term vector $y(t-i) \in \mathbb{R}^n$ and $f_k(u(t-j)) \in \mathbb{R}^m$ belongs to I_c, then the corresponding i, j and k are selected as the elements of I_{ys}, I_{us} and I_{fs} respectively. For example, if the significant terms of $y_w(t-i)$ in I_c look like those in Figure 4.5, then $I_{ys} = \{2,4,6\}$. This rule is relatively conservative because a whole column will be selected if any element in that column is significant. However, it is used here because preserving a small set of redundant terms is often better and also more reasonable than deleting some significant terms.

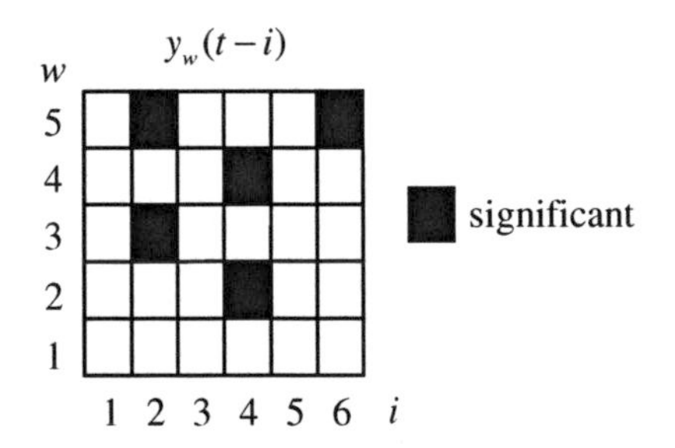

Fig. 4.5 Structure design of Hammerstein model

As a result, the model structure of linear and nonlinear parts of the Hammerstein model can be determined as follows

$$y(t) = A_s(q^{-1})y(t) + B_s(q^{-1})v(t) \text{ ,} \tag{4.24}$$

$$A_s(q^{-1}) = A_{i_1} q^{-i_1} + \cdots + A_{i_{n_{ys}}} q^{-i_{n_{ys}}} \text{ ,} \tag{4.25}$$

$$B_s(q^{-1}) = B_{j_1} q^{-j_1} + \cdots + B_{j_{n_{us}}} q^{-j_{n_{us}}} \text{ ,} \tag{4.26}$$

$$v(t) = F_s(u) = C_{k_1} f_{k_1}(u) + \cdots + C_{k_{n_{fs}}} f_{k_{n_{fs}}}(u) . \tag{4.27}$$

4.5.3 Parameter Estimation

Substituting (4.25)-(4.27) into (4.24) will have

$$y(t) = \sum_{r=1}^{n_{ys}} A_{i_r} y(t - i_r) + \sum_{w=1}^{n_{us}} B_{j_w} \sum_{v=1}^{n_{fs}} C_{k_v} f_{k_v}(u(t - j_w)) + e(t) . \tag{4.28}$$

The problem is to estimate unknown parameter matrices A_{i_r} $(r = 1,...,n_{ys})$, B_{j_w} $(w = 1,...,n_{us})$ and C_{k_v} $(v = 1,...,n_{fs})$ from the data set $\{u(t), y(t)\}_{t=1}^{L}$.

Least-squares Estimation

Define $D_{j_w k_v} = B_{j_w} C_{k_v} \in \mathbb{R}^{n \times m}$, then (4.28) can be rewritten as

$$y(t) = \sum_{r=1}^{n_{ys}} A_{i_r} y(t - i_r) + \sum_{w=1}^{n_{us}} \sum_{v=1}^{n_{fs}} D_{j_w k_v} f_{k_v}(u(t - j_w)) + e(t) . \tag{4.29}$$

Obviously, (4.29) can be expressed as a linear regression form

$$y(t) = \Theta \Phi(t) + e(t) , \tag{4.30}$$

where

$$\Theta = [A_{i_1}, ..., A_{i_{n_{ys}}}, D_{j_1 k_1}, ..., D_{j_1 k_{n_{fs}}}, ..., D_{j_{n_{us}} k_{n_{fs}}}] \in \mathbb{R}^{n \times (n n_{ys} + m n_{us} n_{fs})} , \tag{4.31}$$

$$\Phi(t) = [y(t - i_1)^T, ..., y(t - i_{n_{ys}})^T, f_{k_1}(u(t - j_1))^T, ..., f_{k_{n_{fs}}}(u(t - j_{n_{us}}))^T]^T \in \mathbb{R}^{n n_{ys} + m n_{us} n_{fs}} . \tag{4.32}$$

It is well known that the estimate $\hat{\Theta}$ can be obtained using the recursive least-squares method. Then $\hat{A}_{i_r}$ (r=1,..., n_{ys}) and $\hat{D}_{j_w k_v}$ (w=1,..., n_{us}, v=1,..., n_{fs}) can be easily derived from $\hat{\Theta}$. Now the problem is to reconstruct B_{j_w} and C_{k_v} from $\hat{D}_{j_w k_v}$.

Singular Value Decomposition

With the following definitions for the matrices B and C,

$$B = [B_{j_1}^T, B_{j_2}^T, ..., B_{j_{n_{us}}}^T]^T \in \mathbb{R}^{n n_{us} \times m} , \tag{4.33}$$

$$C = [C_{k_1}, C_{k_2}, ..., C_{k_{n_{fs}}}]^T \in \mathbb{R}^{m n_{fs} \times m} , \tag{4.34}$$

it is easy to see that

$$BC^T = \begin{bmatrix} B_{j_1}C_{k_1} & \cdots & B_{j_1}C_{k_{nfs}} \\ B_{j_2}C_{k_1} & \cdots & B_{j_2}C_{k_{nfs}} \\ \vdots & \ddots & \vdots \\ B_{j_{nus}}C_{k_1} & \cdots & B_{j_{nus}}C_{k_{nfs}} \end{bmatrix} = D \in \mathbb{R}^{nn_{us} \times mn_{fs}} . \tag{4.35}$$

An estimate $\hat{D}$ of the matrix D can then be obtained from the estimate $\hat{D}_{j_w k_v}$. The problem now is how to estimate the parameter matrices B and C from the estimate $\hat{D}$. It is clear that the closest, in the Frobenius norm sense, estimates $\hat{B}$ and $\hat{C}$ are those that solve the following optimization problem

$$(\hat{B}, \hat{C}) = \arg\min_{B,C} \{ \| \hat{D} - BC^T \|_F^2 \} .$$

The solution to this optimization problem is provided by the SVD of the matrix $\hat{D}$ (Bai, 1998; Gómez & Baeyens, 2004).

Theorem 4.1 (Gómez & Baeyens, 2004):

Let $\hat{D} \in \mathbb{R}^{nn_{us} \times mn_{fs}}$ have rank $\gamma \geq m$, and let the economy-size SVD of $\hat{D}$ be given by

$$\hat{D} = U_\gamma \Sigma_\gamma V_\gamma^T = \sum_{i=1}^{\gamma} \sigma_i \mu_i \upsilon_i^T , \tag{4.36}$$

where the singular matrix $\Sigma_\gamma = diag\{\sigma_i\}$ such that

$$\sigma_1 \geq \cdots \geq \sigma_\gamma > 0 ,$$

and where the matrices $U_\gamma = [\mu_1, \ldots, \mu_\gamma] \in \mathbb{R}^{nn_{us} \times \gamma}$ and $V_\gamma = [\upsilon_1, \ldots, \upsilon_\gamma] \in \mathbb{R}^{mn_{fs} \times \gamma}$ contain only the first γ columns of the unitary matrices $U \in \mathbb{R}^{nn_{us} \times nn_{us}}$ and $V \in \mathbb{R}^{mn_{fs} \times mn_{fs}}$ provided by the full SVD of $\hat{D}$,

$$\hat{D} = U \Sigma V^T , \tag{4.37}$$

respectively. Then the matrices $\hat{B} \in \mathbb{R}^{nn_{us} \times m}$ and $\hat{C} \in \mathbb{R}^{mn_{fs} \times m}$ that minimize the norm $\| \hat{D} - BC^T \|_F^2$, are given by

$$(\hat{B}, \hat{C}) = \arg\min_{B,C} \{ \| \hat{D} - BC^T \|_F^2 \} = (U_1, V_1 \Sigma_1) , \tag{4.38}$$

where $U_1 \in \mathbb{R}^{nn_{us} \times m}$, $V_1 \in \mathbb{R}^{mn_{fs} \times m}$ and $\Sigma_1 = diag\{\sigma_1, \ldots, \sigma_m\}$ are given by the following partition of the economy-size SVD in (4.36)

$$\hat{D} = [U_1 \quad U_2] \begin{bmatrix} \Sigma_1 & 0 \\ 0 & \Sigma_2 \end{bmatrix} \begin{bmatrix} V_1^T \\ V_2^T \end{bmatrix} , \tag{4.39}$$

and the approximation error is given by

$$\| \hat{D} - \hat{B}\hat{C}^T \|_F^2 = \sum_{i=m+1}^{\gamma} \sigma_i^2 . \tag{4.40}$$

The identification algorithm can then be summarized as follows. ■

Algorithm 4.1:

Step 1: Use the measured output $\{y(x_i,t)\}_{i=1,t=1}^{N,L}$ as snapshots, find the spatial basis functions $\{\varphi_i(x)\}_{i=1}^n$ via Karhunen-Loève decomposition, and calculate the temporal coefficients $\{y(t)\}_{t=1}^L$ using (4.5).

Step 2: Determine the significant terms of the Hammerstein model as in (4.24) and (4.27) using the orthogonal forward regression in Section 4.5.2.

Step 3: Compute the least-squares estimate $\hat{\Theta}$ in (4.30), and then obtain $\hat{A}_{i_r}$ (r=1,..., n_{ys}) and $\hat{D}_{j_w k_v}$ (w=1,..., n_{us}, v=1,..., n_{fs}) from $\hat{\Theta}$ as in (4.31).

Step 4: Construct the matrix $\hat{D}$ using $\hat{D}_{j_w k_v}$ as in (4.35), and then compute the economy-size SVD of $\hat{D}$ as in (4.36), and the partition of this decomposition as in (4.39).

Step 5: Compute $\hat{B}$ and $\hat{C}$ as $\hat{B} = U_1$ and $\hat{C} = V_1\Sigma_1$ respectively, and then obtain the estimates of the parameter matrices $\hat{B}_{j_w}$ $(w = 1,...,n_{us})$ and $\hat{C}_{k_v}$ $(v = 1,...,n_{fs})$ from $\hat{B}$ and $\hat{C}$ as in (4.33) and (4.34). ■

Finally, the estimated Hammerstein model can be used in the simulation mode for the spatio-temporal dynamics prediction as below

$$\hat{y}(t) = \sum_{r=1}^{n_{ys}} \hat{A}_{i_r} \hat{y}(t - i_r) + + \sum_{w=1}^{n_{us}} \hat{B}_{j_w} \sum_{v=1}^{n_{fs}} \hat{C}_{k_v} f_{k_v}(u(t - j_w)) ,$$

$$\hat{y}_n(x,t) = \sum_{i=1}^{n} \varphi_i(x) \hat{y}_i(t) .$$

4.6 Simulation and Experiment

In order to evaluate the presented modeling method, firstly the simulation on a typical distributed processes: the catalytic rod is given. Then we apply it to the snap curing oven.

The two models to be compared are stated as follows:

- Karhunen-Loève based Hammerstein (KL-Hammerstein) model,
- Spline functions based Hammerstein (SP-Hammerstein) model.

The SP-Hammerstein model is constructed by replacing Karhunen-Loève basis functions φ in (4.5) with spline functions during the modeling procedure. See the reference (Shikin & Plis, 1995; Coca & Billings, 2002) for details on the construction of spline functions.

Define $y(x,t)$ and $\hat{y}_n(x,t)$ as the measured output and the prediction output respectively. Some performance indexes are set up for an easy comparison as follows:

- Spatio-temporal error, $e(x,t) = y(x,t) - \hat{y}_n(x,t)$,
- Spatial normalized absolute error, $SNAE(t) = \frac{1}{N}\sum_{i=1}^{N} | e(x_i,t) |$,
- Temporal normalized absolute error, $TNAE(x) = \sum | e(x,t) | / \sum \Delta t$.

4.6.1 Catalytic Rod

Consider the catalytic rod given in Sections 1.1.2 and 3.6.1. In the simulation, assume the process noise $d(x,t)$ in (3.31) is zero. The inputs are $u_i(t) = 1.1 + 5\sin(t/10 + i/10)$ ($i = 1,...,4$). Nineteen sensors uniformly distributed in the space are used for measurements. The sampling interval Δt is 0.01 and the simulation time is 5. Totally the 500 data are collected, where the first 300 data are used for model estimation, the next 100 data for validation, and the remaining 100 data for model testing.

The measured output $y(x,t)$ of the system is shown in Figure 4.6. As shown in Figure 4.7, the first five Karhunen-Loève basis functions are used for the KL-Hammerstein modeling. Using the cross-validation, the parameters for the linear part of the Hammerstein model are set to $n_y = 3$ and $n_u = 9$. The radial basis functions $f_k(u) = \exp\{-\| u - c_k \|_2^2 / 2\sigma^2\}$ ($k = 1,...,n_f$, $n_f = 10$) with the centers c_k uniformly distributed in the (-3.9, 6.1) and the width $\sigma = 1$ are selected as the basis functions of the nonlinear part. Starting with this initial model, the OFR algorithm in Section 4.5.2 leads to the following compact Hammerstein model

$$y(t) = A_s(q^{-1})y(t) + B_s(q^{-1})v(t),$$

$$A_s(q^{-1}) = A_1q^{-1} + A_2q^{-2} + A_3q^{-3},$$

$$B_s(q^{-1}) = B_1q^{-1} + B_2q^{-2} + B_4q^{-4} + B_9q^{-9},$$

$$v = F_s(u) = C_1f_1(u) + C_2f_2(u) + C_4f_4(u) + C_5f_5(u) + C_6f_6(u) + C_7f_7(u),$$

where the unknown parameters are estimated using the LSE-SVD algorithm.

The predicted output $\hat{y}_n(x,t)$ and prediction error $e(x,t)$ over the whole data set of KL-Hammerstein model are presented in Figure 4.8 and Figure 4.9 respectively. Obviously the KL-Hammerstein model can approximate the spatio-temporal dynamics of original system very well. Now the performance of KL-Hammerstein model is compared with SP-Hammerstein model. As shown in Figure 4.10, eleven third-order splines are used as spatial basis functions in the SP-Hammerstein modeling. Figure 4.11 displays *SNAE*(t) of these two models over the whole data set, where the solid line corresponds to KL-Hammerstein model and the dashed line to SP-Hammerstein model. It can be found that the performance of KL-Hammerstein model is much better than SP-Hammerstein model even if SP-Hammerstein model uses more number of basis functions. This is owing to the optimal Karhunen-Loève basis functions. The KL-Hammerstein model is very efficient for this nonlinear distributed parameter process. For the KL-Hammerstein modeling, two algorithms: OFR-LSE-SVD and LSE-SVD (i.e., without the structure selection using the OFR algorithm) are also compared. The OFR algorithm can make the KL-Hammerstein model compact. In fact, the simulation in Figure 4.12 shows that it can also obtain a more accurate model because a suitable model structure can be selected.

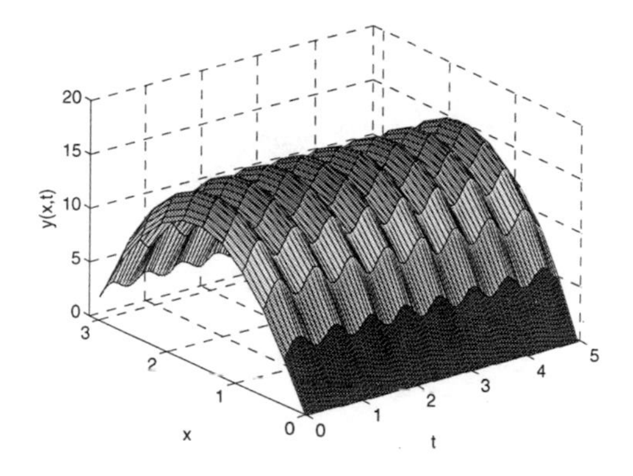

Fig. 4.6 Catalytic rod: Measured output for Hammerstein modeling

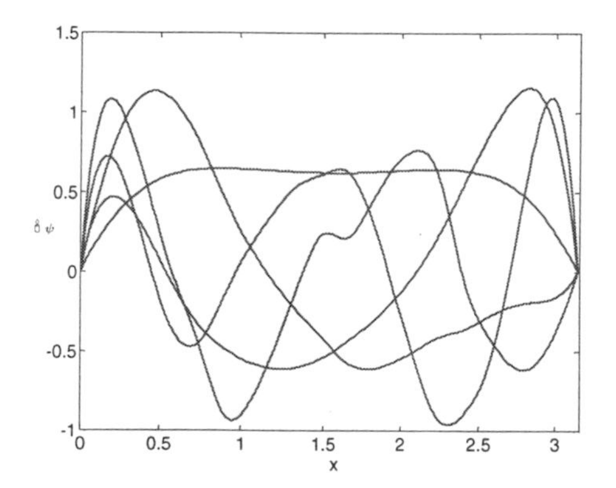

Fig. 4.7 Catalytic rod: KL basis functions for KL-Hammerstein modeling

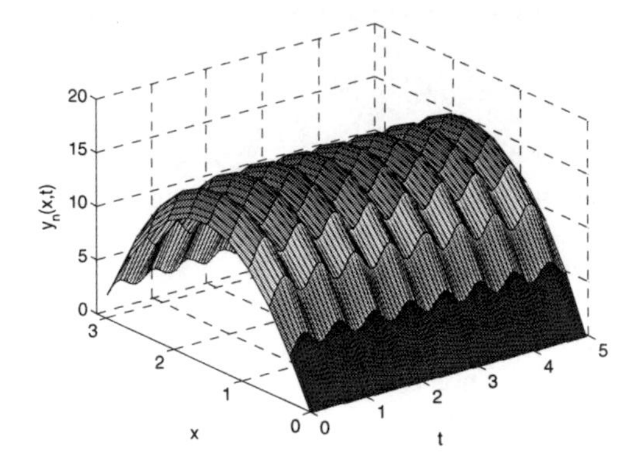

Fig. 4.8 Catalytic rod: KL-Hammerstein model output

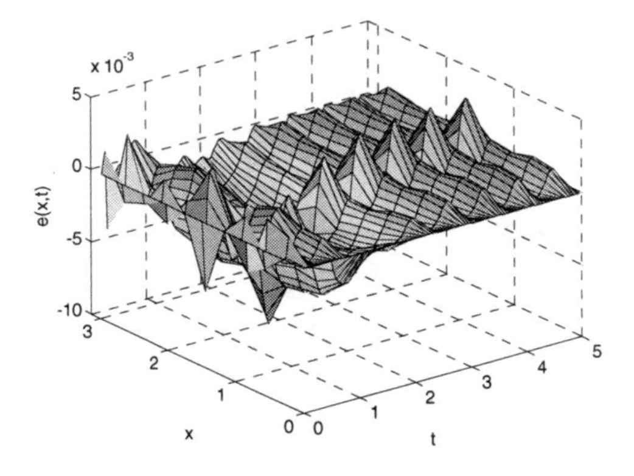

Fig. 4.9 Catalytic rod: Prediction error of KL-Hammerstein model

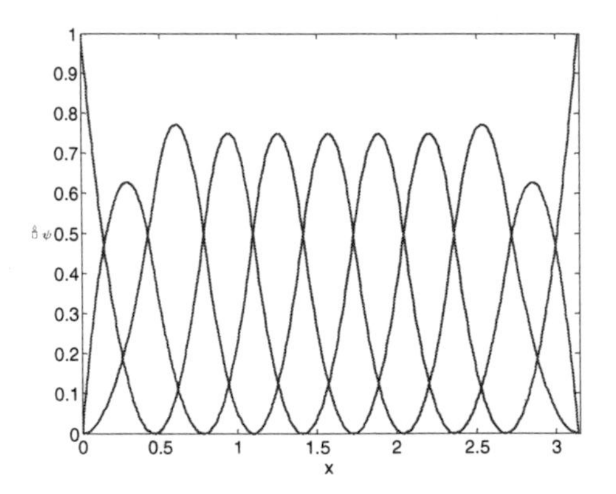

Fig. 4.10 Catalytic rod: Spline basis functions for SP-Hammerstein modeling

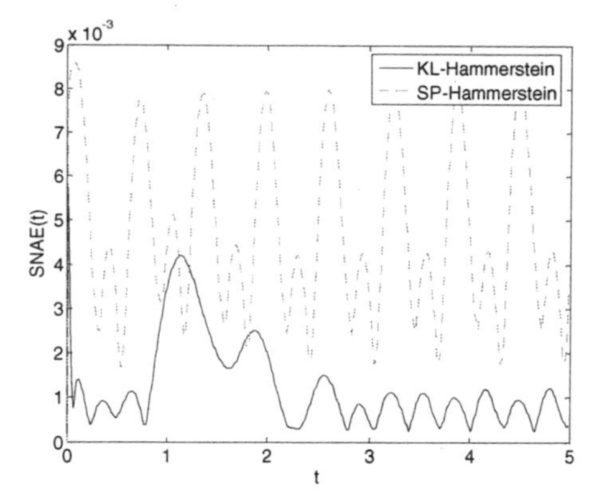

Fig. 4.11 Catalytic rod: Comparison of SP- and KL-Hammerstein models

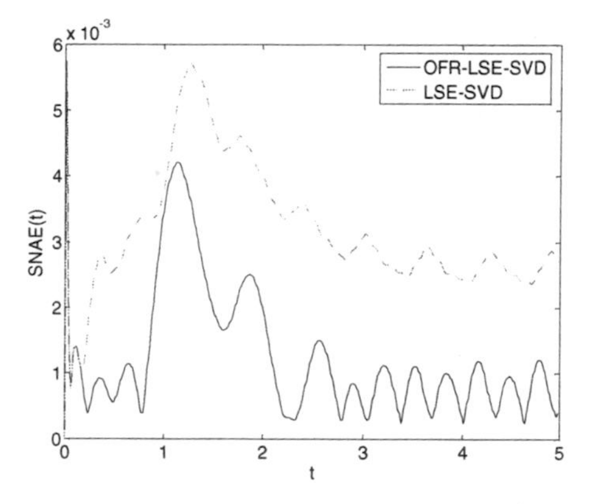

Fig. 4.12 Catalytic rod: Comparison of OFR-LSE-SVD and LSE-SVD algorithms for KL-Hammerstein model

4.6.2 Snap Curing Oven

Consider the snap curing oven (Figure 1.1 and Figure 3.11) provided in Sections 1.1.1 and 3.6.2. In the experiment, a total of 2100 measurements are collected with a sampling interval $\Delta t = 10$ seconds. One thousand and four hundred of measurements from sensors (s1-s5, s7-s10, and s12-s16) are used to estimate the model. The last 700 measurements from the sensors (s1-s5, s7-s10, and s12-s16) are chosen to validate the model during the training. All 2100 measurements from the rest sensors (s6, s11) are used for model testing.

In the *KL-Hammerstein model* modeling, five two-dimensional Karhunen-Loève basis functions are used as spatial bases and the first two of them are shown in Figure 4.13 and Figure 4.14. The parameters for the linear part of the model are $n_y = 6$ and $n_u = 1$. The basis functions $f_k(u)$ of the nonlinear part are designed as standard polynomials $f_k(u) = u^k$ ($k = 1,...,n_f$, $n_f = 2$). From this initial model structure, the significant term selection procedure using the OFR algorithm leads to the following parsimonious model

$$y(t) = A_s(q^{-1})y(t) + B_s(q^{-1})v(t),$$

$$A_s(q^{-1}) = A_1q^{-1} + A_2q^{-2} + A_3q^{-3} + A_5q^{-5},$$

$$B_s(q^{-1}) = B_1q^{-1},$$

$$v = F_s(u) = C_1f_1(u) + C_2f_2(u).$$

After the parameters are estimated using the first 1400 data from the sensors (s1-s5, s7-s10, and s12-s16), the Hammerstein model can be obtained with the significant performance such as the sensor s1 in Figure 4.15. It also performs very well for the untrained locations such as the sensor s6 in Figure 4.16. The predicted temperature distribution of the oven at t=10000s is provided in Figure 4.17.

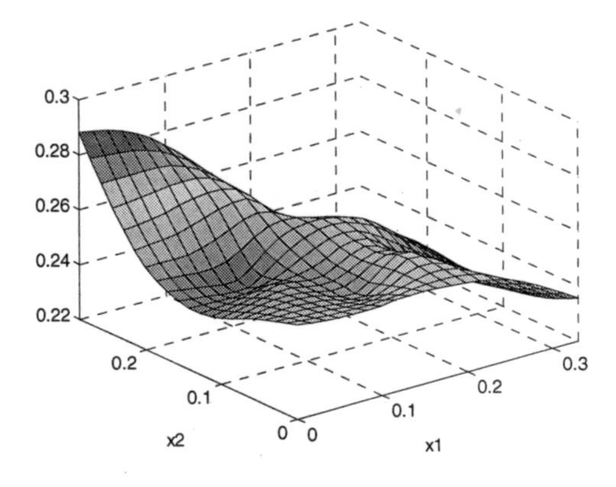

Fig. 4.13 Snap curing oven: KL basis functions (i=1) for KL-Hammerstein modeling

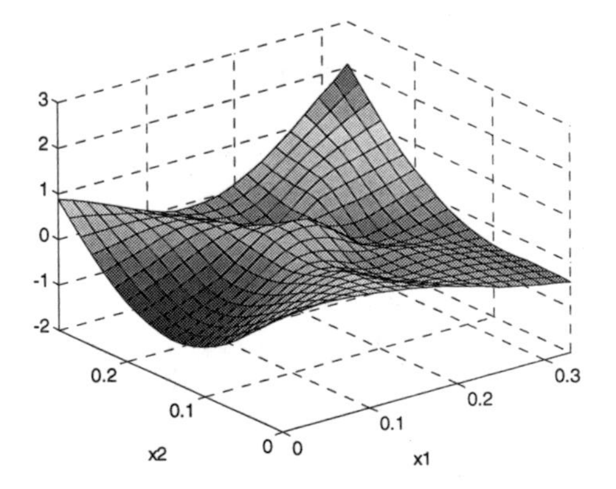

Fig. 4.14 Snap curing oven: KL basis functions (i=2) for KL-Hammerstein modeling

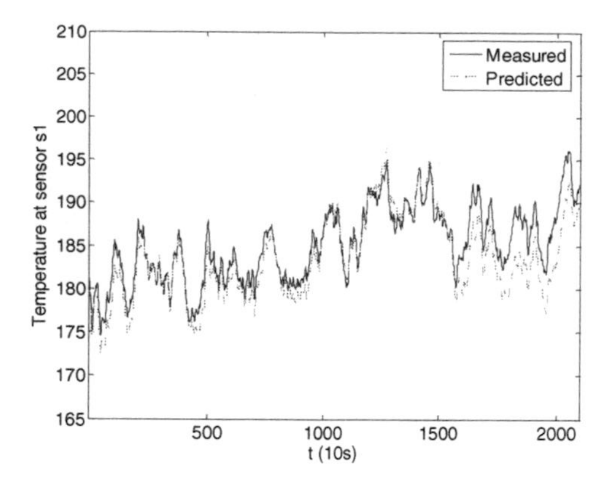

Fig. 4.15 Snap curing oven: Performance of KL-Hammerstein model at sensor s1

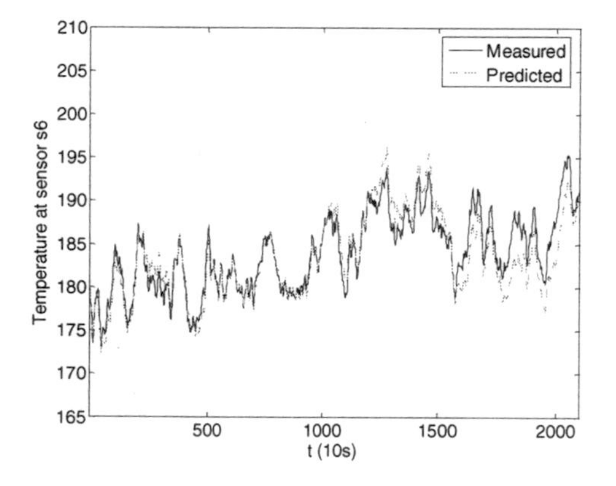

Fig. 4.16 Snap curing oven: Performance of KL-Hammerstein model at sensor s6

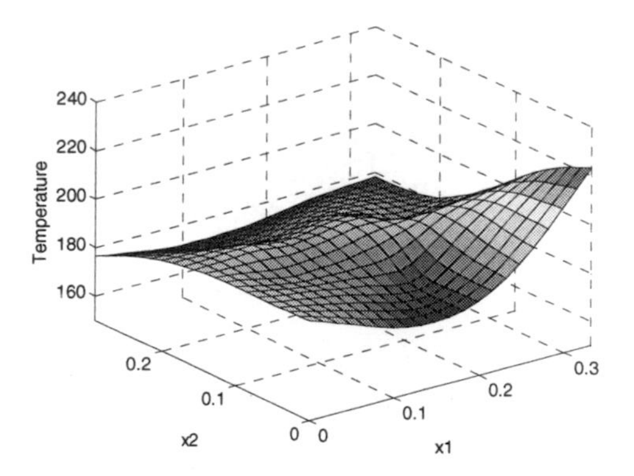

Fig. 4.17 Snap curing oven: Predicted temperature distribution of KL-Hammerstein model at t=10000s

In order to provide a comparison, a SP-Hammerstein model is also constructed using nine third-order splines as spatial basis functions. The first two of them are shown in Figure 4.18 and Figure 4.19. The performance index *TNAE*(x) over the whole data set in Table 4.1 shows that the KL-Hammerstein model works much better than the SP-Hammerstein model because of local spline basis functions used in the SP-Hammerstein model. As shown in Table 4.2, the OFR-LSE-SVD algorithm can produce a more accurate KL-Hammerstein model than the LSE-SVD algorithm. The effectiveness of the presented modeling method is clearly demonstrated in this real application.

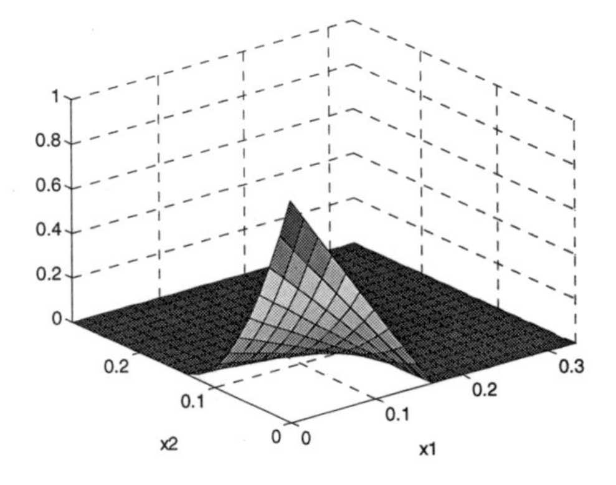

Fig. 4.18 Snap curing oven: Spline basis functions (i=1) for SP-Hammerstein modeling

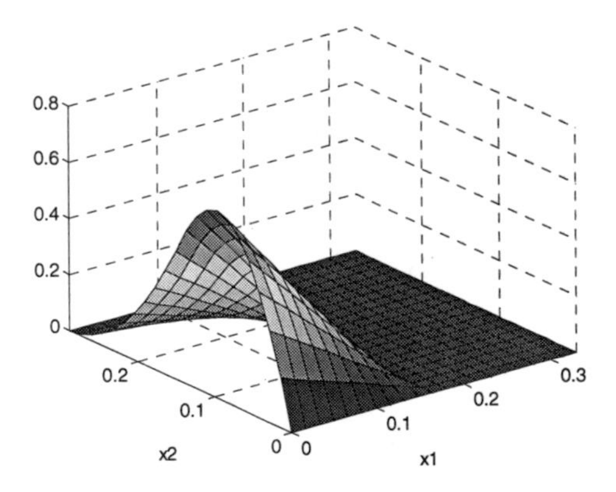

Fig. 4.19 Snap curing oven: Spline basis functions (i=2) for SP-Hammerstein modeling

Table 4.1 Snap curing oven: $TNAE(x)$ of KL-Hammerstein and SP-Hammerstein models

	s1	s2	s3	s4	s5	s6	s7	s8
KL-Hammerstein	1.88	1.63	2	1.52	2.4	1.52	1.62	1.64
SP-Hammerstein	1.5	3.46	4.1	1.77	2.7	2.86	1.85	2.03
	s9	s10	s11	s12	s13	s14	s15	s16
KL-Hammerstein	1.45	1.83	2.08	1.94	1.57	2.15	1.47	1.54
SP-Hammerstein	4.32	2.81	3.14	4.85	2.4	3.84	3.34	1.2

Table 4.2 Snap curing oven: $TNAE(x)$ of OFR-LSE-SVD and LSE-SVD algorithms for KL-Hammerstein model

	s1	s2	s3	s4	s5	s6	s7	s8
OFR-LSE-SVD	1.88	1.63	2	1.52	2.4	1.52	1.62	1.64
LSE-SVD	2.33	1.76	2.54	1.69	3.09	1.68	1.97	1.43
	s9	s10	s11	s12	s13	s14	s15	s16
OFR-LSE-SVD	1.45	1.83	2.08	1.94	1.57	2.15	1.47	1.54
LSE-SVD	1.46	2.16	2.54	2.52	1.82	2.67	1.67	1.71

4.7 Summary

In this chapter, a KL based Hammerstein modeling approach is presented for unknown nonlinear distributed parameter processes with the spatio-temporal output. A Hammerstein distributed parameter system is presented with a static nonlinearity followed by a linear DPS. After the time/space separation, this Hammerstein distributed parameter system can be represented by the traditional Hammerstein system with a set of spatial basis functions. The time/space separation is implemented using the Karhunen-Loève method, where the spatio-temporal output is expanded onto a small number of dominant spatial basis functions with temporal coefficients. Then a low-order compact Hammerstein model is estimated from the low-dimensional temporal data, where the compact model structure is selected based on the orthogonal forward regression, and the parameters are estimated using the least-squares estimation and the singular value decomposition. The algorithm does not require any nonlinear optimization and is numerically robust. The simulation and experiment are presented to show the effectiveness of this spatio-temporal modeling method.

References

1. Bai, E.W.: An optimal two-stage identification algorithm for Hammerstein-Wiener nonlinear systems. Automatica 34(3), 333–338 (1998)
2. Billings, S.A., Korenberg, M.J., Chen, S.: Identification of output-affine systems using an orthogonal least squares algorithm. International Journal of Systems Science 19(8), 1559–1568 (1988a)

3. Billings, S.A., Korenberg, M.J., Chen, S.: Identification of MIMO non-linear systems using a forward-regression orthogonal estimator. International Journal of Control 49(6), 2157–2189 (1988b)
4. Chen, H.F.: Pathwise convergence of recursive identification algorithms for Hammerstein systems. IEEE Transactions on Automatic Control 49(10), 1641–1649 (2004)
5. Coca, D., Billings, S.A.: Identification of finite dimensional models of infinite dimensional dynamical systems. Automatica 38(11), 1851–1865 (2002)
6. Fruzzetti, K.P., Palazoglu, A., McDonald, K.A.: Nonlinear model predictive control using Hammerstein models. Journal of Process Control 7(1), 31–41 (1997)
7. Gómez, J.C., Baeyens, E.: Identification of block-oriented nonlinear systems using orthonormal bases. Journal of Process Control 14(6), 685–697 (2004)
8. Greblicki, W.: Continuous-time Hammerstein system identification from sampled data. IEEE Transactions on Automatic Control 51(7), 1195–1200 (2006)
9. Haber, R., Unbehauen, H.: Structure identification of nonlinear dynamic systems - A survey on input/output approaches. Automatica 26(4), 651–677 (1990)
10. Lind, I., Ljung, L.: Regressor and structure selection in NARX models using a structured ANOVA approach. Automatica 44(2), 383–395 (2008)
11. Narendra, K., Gallman, P.: An iterative method for the identification of nonlinear systems using a Hammerstein model. IEEE Transactions on Automatic Control 11(3), 546–550 (1966)
12. Piroddi, L., Spinelli, W.: An identification algorithm for polynomial NARX models based on simulation error minimization. International Journal of Control 76(17), 1767–1781 (2003)
13. Samuelsson, P., Norlander, H., Carlsson, B.: An integrating linearization method for Hammerstein models. Automatica 41(10), 1825–1828 (2005)
14. Shikin, E.V., Plis, A.I.: Handbook on splines for the user. CRC Press, Boca Raton (1995)
15. Sjöberg, J., Zhang, Q., Ljung, L., Benveniste, A., Delyon, B., Glorennec, P., Hjalmarsson, H., Juditsky, A.: Nonlinear black-box modeling in system identification: A unified approach. Automatica 31(12), 1691–1724 (1995)
16. Stoica, P., Söderström, T.: Instrumental-variable methods for identification of Hammerstein systems. International Journal of Control 35(3), 459–476 (1982)
17. Vörös, J.: Recursive identification of Hammerstein systems with discontinuous nonlinearities containing dead-zones. IEEE Transactions on Automatic Control 48(12), 2203–2206 (2003)
18. Zhu, Y.C.: Identification of Hammerstein models for control using ASYM. International Journal of Control 73(18), 1692–1702 (2000)

5 Multi-channel Spatio-Temporal Modeling for Hammerstein Distributed Parameter Systems

Abstract. A multi-channel spatio-temporal Hammerstein modeling approach is presented in this chapter. As a special case of the model described in Chapter 4, a spatio-temporal Hammerstein model is constructed with a static nonlinearity followed by a linear spatio-temporal kernel. When the model structure is matched with the system, a basic single-channel identification algorithm with the algorithm used in the Chapter 4 can work well. When there is unmodeled dynamics, a multi-channel modeling framework can provide a better performance, because more channels used can attract more information from the process. The modeling convergence can be guaranteed under noisy measurements. The simulation example and the experiment on snap curing oven are presented to show the effectiveness of this modeling method.

5.1 Introduction

In the process control, Hammerstein models have been successfully used to represent many practical nonlinear ODE processes (Eskinat, Johnson & Luyben, 1991; Fruzzetti, Palazoglu & McDonald, 1997). However, Hammerstein models are only studied for lumped parameter systems (LPS) because they are only temporal models and can not model spatial dynamics. This chapter will extend the traditional Hammerstein modeling into nonlinear distributed parameter systems (DPS) via the spatio-temporal kernel idea.

For the traditional Hammerstein modeling, several methods have been proposed in the literature (Narendra & Gallman, 1966; Stoica, 1981; Bai, 1998; Bai & Li, 2004; Chen, 2004; Zhu, 2000; Greblicki, 2006; Vörös, 2003; Gómez & Baeyens, 2004). It is notable that an algorithm based on the least-squares estimation and the singular value decomposition (LSE-SVD) is proposed for Hammerstein-Wiener systems (Bai, 1998) and extensively studied for Hammerstein systems (Gómez & Baeyens, 2004). The algorithm is derived from the use of basis functions for the representation of the linear and the nonlinear parts. In the case of model matching, the consistency of the estimates can be guaranteed under certain conditions. However, in the presence of unmodeled dynamics, further studies are required.

In this chapter, in order to model the nonlinear distributed parameter system, a spatio-temporal Hammerstein model is presented by adding the space variables into the traditional Hammerstein model, which consists of the cascade connection of a static nonlinearity followed by a distributed dynamical linear time-invariant system.

H.-X. Li and C. Qi: Spatio-Temporal Modeling of Nonlinear DPS, ISCA 50, pp. 95–121.
springerlink.com

The linear time-invariant DPS is represented by a spatio-temporal kernel, i.e., Green's function in Section 2.5.1. This spatio-temporal kernel makes the Hammerstein model have spatio-temporal modeling capability.

A basic identification algorithm based on LSE-SVD is designed as follows. Firstly, the nonlinear and the distributed linear parts are expanded onto spatial and temporal basis functions with unknown coefficients. In order to reduce the parametric complexity, the Karhunen-Loève (KL) decomposition is used to find the dominant spatial basis functions and Laguerre polynomials are selected as the temporal basis functions. Then, using the Galerkin method, the spatio-temporal modeling will turn into a traditional modeling problem in the time domain. Subsequently, the least-squares techniques can be used to identify a parameter matrix charactering the product of parameters of the linear and the nonlinear parts. Finally, by using SVD, optimal estimates of the parameters of each part can be obtained. This basic identification algorithm can provide consistent estimates under some assumptions in the case of model matching.

In the presence of unmodeled dynamics, a multi-channel identification algorithm is presented to compensate the residuals of the single-channel model and further reduce the modeling error. This algorithm is noniterative and numerically robust since it is based only on the least-squares estimation and the singular value decomposition. The convergent estimates can be guaranteed under proper assumptions. The spatio-temporal Hammerstein model can be easily used for many applications such as model predictive control due to its simple nonlinear structure. The simulation and experiment demonstrate the effectiveness of the presented modeling method.

The difference with Chapter 4 is described as follows:

- Different Hammerstein models are used. In this chapter, a new Hammerstein model is constructed with a Green's function (time/space nature) and a static nonlinear function. This new constructed Hammerstein model has time/space nature and is used to directly model DPS. It can be considered as a kernel-based scheme. In Chapter 4, the Hammerstein distributed parameter system is constructed from a lumped Hammerstein system via the time/space synthesis based on the Karhunen-Loève method.
- Different identification algorithms are used. In this chapter, the kernel structure of the multi-channel Hammerstein system is given, so the identification mainly focuses on the parameter estimation with the LSE-SVD algorithm. In Chapter 4, the modeling is a single-channel Hammerstein modeling, and the identification needs to consider both model structure design and parameter estimation with the OFR and the LSE-SVD algorithm.

This chapter is organized as follows. In Section 5.2, the Hammerstein distributed parameter system is presented via the spatio-temporal kernel. The single-channel identification algorithm is derived in Section 5.3. The multi-channel modeling approach and analysis are provided in Section 5.4. Simulation example and experiment are presented to illustrate the performance of the presented modeling approach in Section 5.5, and finally, some conclusions are provided in Section 5.6.

5.2 Hammerstein Distributed Parameter System

A Hammerstein distributed parameter system is shown in Figure 5.1. The system consists of a static nonlinear element $N(\cdot):\mathbb{R}\to\mathbb{R}$ followed by a distributed linear time-invariant system

$$y(x,t)=\sum_{\tau=0}^{t}\int_{\Omega}g(x,\zeta,\tau)v(\zeta,t-\tau)d\zeta\ , \tag{5.1}$$

Here a spatio-temporal kernel model, i.e., Green's function Θ^* is used to represent the linear DPS with a transfer function $G(x,\zeta,q)$ (1×1), where x and ζ are spatial variables defined on the domain Ω, and q stands for the forward shift operator. The input-output relationship of the system is then given by

$$y(x,t)=\sum_{\tau=0}^{t}\int_{\Omega}g(x,\zeta,\tau)N(u(\zeta,t-\tau))d\zeta+d(x,t)\ , \tag{5.2}$$

where $u(x,t)\in\mathbb{R}$ and $y(x,t)\in\mathbb{R}$ are the input and output at time t, and $d(x,t)\in\mathbb{R}$ includes the unmodeled dynamics and the stochastic disturbance. For easy understanding, the integral operator is used for spatial operation and sum operator for temporal operation. In this study, only the single-input-single-output (SISO) system is considered. The extension of the results to the multi-input-multi-output (MIMO) system is straightforward.

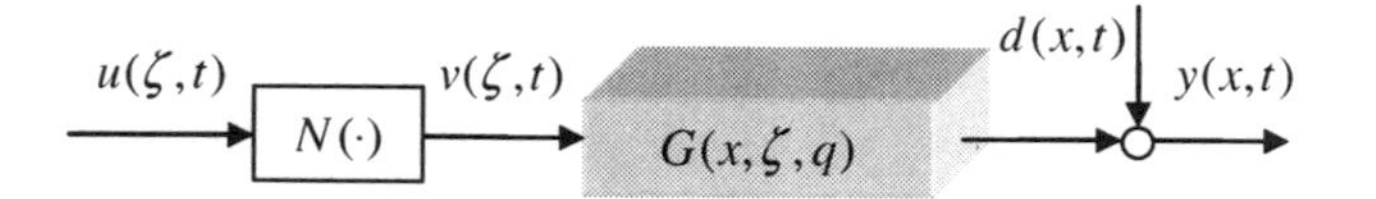

Fig. 5.1 Hammerstein distributed parameter system

The problem is to estimate N and G from the input-output data $\{u(\zeta_i,t),y(x_j,t)\}$, ($i=1,\ldots,n_u$, $j=1,\ldots,n_y$, $t=1,\ldots,n_t$), where $\zeta_i,x_j\in\Omega$, n_u and n_y are the number of sampled spatial points of the input and output, and n_t is the time length. For simplicity, assume that the spatial points ζ_i and x_j are uniformly distributed over the spatial domain.

5.3 Basic Identification Approach

5.3.1 Basis Function Expansion

In general, the input $u(x,t)$ has finite degrees of freedom since only a finite number of actuators are available in practice. Thus assume that the input $u(x,t)$ can be formulated in terms of a finite number of spatial input basis functions $\{\psi_i(x)\}_{i=1}^m$ as follows

$$u(x,t) = \sum_{i=1}^{m} \psi_i(x) a_i(t) , \tag{5.3}$$

where $a_i(t) = \int_\Omega u(x,t)\psi_i(x)dx$ is the time coefficient (implemental input signal), $\psi_i(x)$ describes how the control action $a_i(t)$ is distributed in the spatial domain Ω, and m is the number of actuators, which can be determined by physical knowledge.

Ideally, the output $y(x,t)$ and the error $d(x,t)$ can be expressed by an infinite set of orthonormal spatial output basis functions $\{\varphi_i(x)\}_{i=1}^\infty$ as follows

$$y(x,t) = \sum_{i=1}^{\infty} \varphi_i(x) b_i(t) , \tag{5.4}$$

$$d(x,t) = \sum_{i=1}^{\infty} \varphi_i(x) d_i(t) , \tag{5.5}$$

where $b_i(t) = \int_\Omega y(x,t)\varphi_i(x)dx$ and $d_i(t) = \int_\Omega d(x,t)\varphi_i(x)dx$ are the time coefficients respectively. This is because of inherently infinite-dimensional characteristic of distributed parameter system. Practically, both $y(x,t)$ and $d(x,t)$ can be truncated into n dimensions as below

$$y_n(x,t) = \sum_{i=1}^{n} \varphi_i(x) b_i(t) , \tag{5.6}$$

$$d_n(x,t) = \sum_{i=1}^{n} \varphi_i(x) d_i(t) . \tag{5.7}$$

$\varphi_i(x)$ are usually selected as standard orthonormal functions such as Fourier series, Legendre polynomials, Jacobi polynomials and Chebyshev polynomials (Datta & Mohan, 1995). In this study, the KL decomposition (Park & Cho, 1996a, 1996b) is used to identify the empirical dominant basis functions from the process data. Among all linear expansions, the KL expansion is the most efficient in the sense that for a given approximation error, the number of KL bases required is minimal. Owing to this, the KL decomposition can help to reduce the number of estimated parameters.

Assume that the intermediate output $v(x,t) \in \mathbb{R}$ can be described as

$$v(x,t) = N(u(x,t)) = \sum_{i=1}^{m} \sum_{j=1}^{v} \psi_i(x) \beta_j h_j(a_i(t)) , \tag{5.8}$$

where $h_j(\cdot) : \mathbb{R} \to \mathbb{R}$ (j=1,…,v) are nonlinear basis functions and $\beta_j \in \mathbb{R}$ (j=1,…,v) are coefficients. Typically, the nonlinear functions $h_j(\cdot)$ can be chosen as polynomials, radial basis functions, wavelets (Sjöberg *et al.*, 1995) and so on.

Assuming that the kernel $g(x,\zeta,\tau)$ in (5.2) is absolutely integrable on time domain $[0,\infty)$ at any spatial point x and ζ, which means that the corresponding model is stable, then it can be represented by means of orthonormal temporal basis functions. Theoretically, the kernel is supposed to be expanded onto spatial output bases $\{\varphi_i(x)\}_{i=1}^{\infty}$, spatial input bases $\{\psi_i(x)\}_{i=1}^{m}$ and temporal bases $\{\phi_i(t)\}_{i=1}^{\infty}$ as follows

$$g(x,\zeta,\tau)=\sum_{i=1}^{\infty}\sum_{j=1}^{m}\sum_{k=1}^{\infty}\alpha_{i,j,k}\varphi_i(x)\psi_j(\zeta)\phi_k(\tau), \tag{5.9}$$

where $\alpha_{i,j,k}\in\mathbb{R}$ (i=1,…,∞, j=1,…,m, k=1,…,∞) are constant coefficients of basis functions $\varphi_i(x)\psi_j(\zeta)\phi_k(\tau)$. Practically, a finite-dimensional truncation

$$g_{n,l}(x,\zeta,\tau)=\sum_{i=1}^{n}\sum_{j=1}^{m}\sum_{k=1}^{l}\alpha_{i,j,k}\varphi_i(x)\psi_j(\zeta)\phi_k(\tau), \tag{5.10}$$

is often good enough for a realistic approximation, where n and l are the dimension of output bases and temporal bases respectively. $\phi_i(t)$ can be selected as Laguerre functions (Wahlberg, 1991), Kautz functions (Wahlberg, 1994) and generalized orthonormal basis functions (Heuberger, Van den Hof, & Bosgra, 1995). Here, Laguerre functions are chosen for the development, due to their simplicity and robustness to the choice of sampling period and model order (Wahlberg, 1991). Laguerre function is defined as a functional series (Zervos & Dumont, 1988)

$$\phi_i(t)\triangleq\sqrt{2\xi}\frac{e^{\xi t}}{(i-1)!}\cdot\frac{d^{i-1}}{dt^{i-1}}[t^{i-1}\cdot e^{-2\xi t}], i=1,2,...,\infty, \xi>0, \tag{5.11}$$

where ξ is the time-scaling factor and $t\in[0,\infty)$ is time variable. The Laplace transform of the ith Laguerre function is given by (5.12)

$$\phi_i(s)=\sqrt{2\xi}\frac{(s-\xi)^{i-1}}{(s+\xi)^{i}}, i=1,2,...,\infty, \xi>0. \tag{5.12}$$

Laguerre functions (5.11) and (5.12) form a complete orthonormal basis in the function space $L_2(R_+)$ and $H_2(C_+)$ respectively.

Substitution of (5.4), (5.5), (5.8) and (5.9) into (5.2) with a n-dimensional truncation of output bases will have

$$\begin{aligned}y_n(x,t)=&\sum_{\tau=0}^{t}\int_{\Omega}\sum_{i=1}^{n}\sum_{j=1}^{m}\sum_{k=1}^{\infty}\alpha_{i,j,k}\varphi_i(x)\psi_j(\zeta)\phi_k(\tau)\sum_{r=1}^{m}\sum_{s=1}^{v}\psi_r(\zeta)\beta_s h_s(a_r(t-\tau))d\zeta\\&+d_n(x,t),\end{aligned} \tag{5.13}$$

To make the kernel $g_{n,l}(x,\zeta,\tau)$ explicit, (5.13) can be rewritten as

$$\begin{aligned}y_n(x,t)=&\sum_{\tau=0}^{t}\int_{\Omega}\sum_{i=1}^{n}\sum_{j=1}^{m}\sum_{k=1}^{l}\alpha_{i,j,k}\varphi_i(x)\psi_j(\zeta)\phi_k(\tau)\sum_{r=1}^{m}\sum_{s=1}^{v}\psi_r(\zeta)\beta_s h_s(a_r(t-\tau))d\zeta\\&+\tilde{d}_n(x,t),\end{aligned} \tag{5.14}$$

where

$$\tilde{d}_n(x,t) = \sum_{i=1}^{n} \varphi_i(x)\tilde{d}_i(t) \, ,$$

$$\tilde{d}_i(t) = \sum_{\tau=0}^{t} \int_\Omega \sum_{j=1}^{m} \sum_{k=l+1}^{\infty} \alpha_{i,j,k} \psi_j(\zeta)\phi_k(\tau) \sum_{r=1}^{m} \sum_{s=1}^{v} \psi_r(\zeta)\beta_s h_s(a_r(t-\tau))d\zeta + d_i(t) \, .$$

5.3.2 Temporal Modeling Problem

Equation (5.14) can be further simplified into

$$\sum_{i=1}^{n} \varphi_i(x) b_i(t) = \sum_{i=1}^{n} \varphi_i(x) \sum_{j=1}^{m} \sum_{k=1}^{l} \alpha_{i,j,k} \sum_{r=1}^{m} \sum_{s=1}^{v} \beta_s \psi_{j,r} \ell_{k,s,r}(t) + \sum_{i=1}^{n} \varphi_i(x) \tilde{d}_i(t) \, , \tag{5.15}$$

where

$$\psi_{j,r} = \int_\Omega \psi_j(\zeta)\psi_r(\zeta)d\zeta \, , \tag{5.16}$$

$$\ell_{k,s,r}(t) = \sum_{\tau=0}^{t} \phi_k(\tau) h_s(a_r(t-\tau)) \, , \tag{5.17}$$

Using the Galerkin method (Christofides, 2001b), the projection of (5.15) onto the output basis functions $\varphi_h(x)$ ($h = 1,...,n$) will lead to the following n equations

$$\sum_{i=1}^{n} \int_\Omega \varphi_h(x)\varphi_i(x)dx b_i(t) = \sum_{i=1}^{n} \int_\Omega \varphi_h(x)\varphi_i(x)dx \sum_{j=1}^{m} \sum_{k=1}^{l} \alpha_{i,j,k} \sum_{r=1}^{m} \sum_{s=1}^{v} \beta_s \psi_{j,r} \ell_{k,s,r}(t)$$
$$+ \sum_{i=1}^{n} \int_\Omega \varphi_h(x)\varphi_i(x)dx \tilde{d}_i(t).$$

Since $\{\varphi_i(x)\}_{i=1}^{n}$ are orthonormal, we have

$$b(t) = \sum_{j=1}^{m} \sum_{k=1}^{l} \sum_{s=1}^{v} \alpha_{j,k} \beta_s L_{j,k,s}(t) + \tilde{d}(t) \, , \tag{5.18}$$

where

$$b(t) = [b_1(t), \cdots, b_n(t)]^T \in \mathbb{R}^n \, , \tag{5.19}$$

$$\tilde{d}(t) = [\tilde{d}_1(t), \cdots, \tilde{d}_n(t)]^T \in \mathbb{R}^n \, . \tag{5.20}$$

$$\alpha_{j,k} = [\alpha_{1,j,k}, \cdots, \alpha_{n,j,k}]^T \in \mathbb{R}^n \, , \tag{5.21}$$

$$L_{j,k,s}(t) = \sum_{r=1}^{m} \psi_{j,r} \ell_{k,s,r}(t) \in \mathbb{R} \, , \tag{5.22}$$

5.3.3 Least-Squares Estimation

Equation (5.18) can be expressed in a linear regression form

$$b(t) = \Theta^T \Phi(t) + \tilde{d}(t), \tag{5.23}$$

where

$$\Theta = [\alpha_{1,1}\beta_1, \cdots, \alpha_{1,1}\beta_v, \cdots, \alpha_{m,l}\beta_1, \cdots, \alpha_{m,l}\beta_v]^T \in \mathbb{R}^{n \times mlv}, \tag{5.24}$$

$$\Phi(t) = [L_{1,1,1}(t), \cdots, L_{1,1,v}(t), \cdots, L_{m,l,1}(t), \cdots, L_{m,l,v}(t)]^T \in \mathbb{R}^{mlv}. \tag{5.25}$$

In practice, u and y are uniformly sampled over the spatial domain. In this case, $b(t)$ can be computed from the pointwise data $y(x,t)$ using spline interpolation in the spatial domain. The accurate $a(t) = [a_1(t), \cdots, a_m(t)]^T$ can be obtained from $u(x,t)$ using the inversion operation of a matrix formed by the basis functions provided that $n_u \geq m$. Then, $\Phi(t)$ can be constructed from $a(t)$.

Considering n_t set of temporal data $\{\Phi(t), b(t)\}_{t=1}^{n_t}$, it is well known from (Ljung, 1999) that by minimizing a quadratic criterion on the prediction errors

$$\hat{\Theta} = \arg\min_{\Theta} \{\frac{1}{n_t} \sum_{t=1}^{n_t} \| b(t) - \Theta^T \Phi(t) \|^2 \}, \tag{5.26}$$

Θ can be estimated using the least-squares method as follows

$$\hat{\Theta} = (\frac{1}{n_t} \sum_{t=1}^{n_t} \Phi(t)\Phi^T(t))^{-1} (\frac{1}{n_t} \sum_{t=1}^{n_t} \Phi(t) b^T(t)), \tag{5.27}$$

provided that the indicated inverse exists.

The next problem is how to estimate the parameters $\alpha_{j,k}$ (j=1,…,m, k=1,…,l) and β_s (s=1,…,v) from the estimate $\hat{\Theta}$ in (5.27).

5.3.4 Singular Value Decomposition

For convenience, we define $\alpha = [\alpha_{1,1}^T, \cdots, \alpha_{m,l}^T]^T \in \mathbb{R}^{nml}$ and $\beta = [\beta_1, \cdots, \beta_v]^T \in \mathbb{R}^v$. It is clear that the parameterization (5.8) and (5.9) is not unique, since any parameter vectors $\alpha\sigma$ and $\beta\sigma^{-1}$, under nonzero constant σ, can provide the same input/output equation (5.14). A technique that can be used to obtain uniqueness is to normalize the parameter vectors α (or β), for instance assuming that $\| \beta \|_2 = 1$. Under this assumption, the parameterization in (5.8) and (5.9) is unique.

From the definition of the parameter matrix Θ in (5.24), it is easy to see that

$$\Theta = blockvec(\Theta_{\alpha\beta}),$$

where $blockvec(\Theta_{\alpha\beta})$ is the block column matrix obtained by stacking the block columns of $\Theta_{\alpha\beta}$ on the top of each other, and $\Theta_{\alpha\beta} \in \mathbb{R}^{v \times nml}$ has been defined as

$$\Theta_{\alpha\beta} \triangleq \begin{bmatrix} \beta_1 \alpha_{1,1}^T & \cdots & \beta_1 \alpha_{m,l}^T \\ \beta_2 \alpha_{1,1}^T & \cdots & \beta_2 \alpha_{m,l}^T \\ \vdots & \ddots & \vdots \\ \beta_v \alpha_{1,1}^T & \cdots & \beta_v \alpha_{m,l}^T \end{bmatrix} = \beta\alpha^T . \tag{5.28}$$

Thus an estimate $\hat{\Theta}_{\alpha\beta}$ of the matrix $\Theta_{\alpha\beta}$ can be obtained from the estimate $\hat{\Theta}$ in (5.27). The problem now is to estimate the parameter matrices α and β from $\hat{\Theta}_{\alpha\beta}$.

In order to solve this problem, an important fact should be pointed out. It is clear that the closest, in the Frobenius norm sense, approximation of $\hat{\Theta}_{\alpha\beta}$ is not just a single pair of $\hat{\alpha}$ and $\hat{\beta}$ but a series of pairs $(\hat{\beta}^c, \hat{\alpha}^c)$, $(c = 1,\ldots,p)$ that solve the following optimization problem

$$(\hat{\beta}^c, \hat{\alpha}^c) = \arg\min_{\alpha^c, \beta^c} \{ \| \hat{\Theta}_{\alpha\beta} - \sum_{c=1}^{p} \beta^c (\alpha^c)^T \|_F^2 \} , \tag{5.29}$$

where the Frobenius norm of a matrix $A \in \mathbb{R}^{m \times n}$ is defined as $\| A \|_F = (\sum_{i=1}^{m} \sum_{j=1}^{n} A_{ij}^2)^{1/2}$.

To illustrate this fact, a lemma (Golub & Van Loan, 1989) should be introduced.

Lemma 5.1:

Let $\hat{\Theta}_{\alpha\beta} \in \mathbb{R}^{v \times nml}$ have rank $\gamma \ge 1$, and let the economy-size SVD of $\hat{\Theta}_{\alpha\beta}$ be given by

$$\hat{\Theta}_{\alpha\beta} = U_\gamma \Sigma_\gamma V_\gamma^T = \sum_{i=1}^{\gamma} \sigma_i \mu_i \upsilon_i^T , \tag{5.30}$$

where the singular matrix $\Sigma_\gamma = diag\{\sigma_i\}$ such that

$$\sigma_1 \ge \cdots \ge \sigma_\gamma > 0 ,$$

and where the matrices $U_\gamma = [\mu_1, \ldots, \mu_\gamma] \in \mathbb{R}^{v \times \gamma}$ and $V_\gamma = [\upsilon_1, \ldots, \upsilon_\gamma] \in \mathbb{R}^{nml \times \gamma}$ contain only the first γ columns of the unitary matrices $U \in \mathbb{R}^{v \times v}$ and $V \in \mathbb{R}^{nml \times nml}$ provided by the full SVD of $\hat{\Theta}_{\alpha\beta}$,

$$\hat{\Theta}_{\alpha\beta} = U \Sigma V^T ,$$

respectively. Then $\forall p \le \gamma$, the following equation holds

$$(\hat{\beta}^c,\hat{\alpha}^c)=\arg\min_{\alpha^c,\beta^c}\{\|\hat{\Theta}_{\alpha\beta}-\sum_{c=1}^{p}\beta^c(\alpha^c)^T\|_F^2\}=(\mu_c,\upsilon_c\sigma_c),(c=1,\ldots,p), \tag{5.31}$$

where $(\hat{\beta}^c,\hat{\alpha}^c)$ is defined as the cth channel, and p is the number of channels. The parameter approximation error is given by

$$\varepsilon^p=\|\hat{\Theta}_{\alpha\beta}-\sum_{c=1}^{p}\beta^c(\alpha^c)^T\|_F^2=\sum_{c=p+1}^{\gamma}\sigma_c^2\,. \tag{5.32}$$

■

Based on Lemma 5.1, the estimated parameters $\hat{\alpha}$ and $\hat{\beta}$ can be obtained by

$$(\hat{\beta},\hat{\alpha})=\arg\min_{\alpha,\beta}\{\|\hat{\Theta}_{\alpha\beta}-\beta(\alpha)^T\|_F^2\}=(\mu_1,\upsilon_1\sigma_1)\,. \tag{5.33}$$

The consistency in the previous work (Bai, 1998; Gómez & Baeyens, 2004) can be extended to this basic identification approach under certain conditions (e.g., model matching and zero-mean disturbance). However, if such conditions are not satisfied, it would be difficult to obtain a solution. In the following section, a novel multi-channel identification approach will be presented to provide a better solution.

5.4 Multi-channel Identification Approach

5.4.1 Motivation

Definition 5.1:
The system (5.2) is named as a single-channel Hammerstein system. The multi-channel Hammerstein system is formed by the parallel connection of p single-channel Hammerstein systems. ■

For a single-channel Hammerstein model, we can see from (5.28) that $rank(\Theta_{\alpha\beta})=1$, since $\Theta_{\alpha\beta}$ is the product of a column vector β and a row vector α^T. However, generally speaking, its estimate $\hat{\Theta}_{\alpha\beta}$ from process data can not be exactly expressed as the product of a column vector and a row vector due to unmodeled dynamics and disturbance. That is $rank(\hat{\Theta}_{\alpha\beta})>1$. The unmodeled dynamics refer to the error between the single-channel model and the system, which may be too large. In addition, σ_2/σ_1 (see (5.30)) can not always be small enough to make the parameter approximation error (5.32) acceptable. Thus, it is very necessary to add more channels to compensate the modeling residuals.

5.4.2 Multi-channel Identification

The presented multi-channel identification methodology is shown in Figure 5.2.

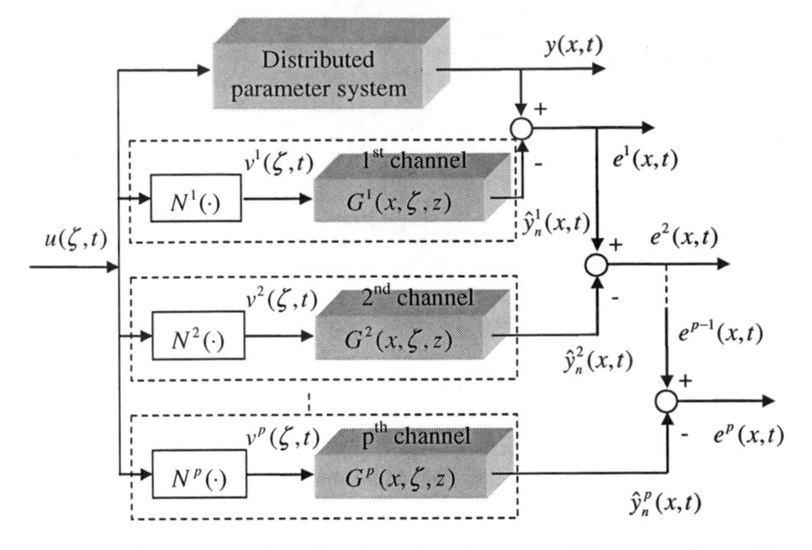

Fig. 5.2 Multi-channel identification of spatio-temporal Hammerstein model

A sequential identification algorithm is designed as follows. Firstly, the 1st channel model is estimated using the basic identification algorithm from the input-output data $\{u(x,t), y(x,t)\}_{t=1}^{n_t}$. Secondly, the 1st channel model error $e^1(x,t) = y(x,t) - \hat{y}_n^1(x,t)$ is regarded as the new output, and then the 2nd channel model is identified. Similarly, $e^2(x,t), \cdots, e^{p-1}(x,t)$ can determine the 3rd, $\cdots$, pth channel models, and so on.

However, the sequential identification algorithm for each channel may lead to a computational burden problem. According to Lemma 5.1, the multi-channel identification algorithm can be easily implemented simultaneously as below.

Algorithm 5.1:

Step 1: Determine the input basis functions $\{\psi_i(x)\}_{i=1}^m$, find the output basis functions $\{\varphi_i(x)\}_{i=1}^n$ using the KL method, choose the Laguerre polynomials $\{\phi_i(t)\}_{i=1}^l$, then obtain the corresponding temporal coefficients $\{a(t)\}_{t=1}^{n_t}$ and $\{b(t)\}_{t=1}^{n_t}$ of the input $\{u(x_i,t)\}_{i=1,t=1}^{n_u,n_t}$ and output $\{y(x_j,t)\}_{j=1,t=1}^{n_y,n_t}$ respectively.

Step 2: Compute the linear regressors $\Phi(t)$ according to (5.17) and (5.25) using $\{a(t)\}_{t=1}^{n_t}$, then compute the least-squares estimate $\hat{\Theta}$ as in (5.27), and construct the matrix $\hat{\Theta}_{\alpha\beta}$ such that $\hat{\Theta} = blockvec(\hat{\Theta}_{\alpha\beta})$.

Step 3: Compute the economy-size SVD of $\hat{\Theta}_{\alpha\beta}$ as in Lemma 5.1, and the partition of this decomposition as in (5.30).

Step 4: Compute the estimates of the parameter vectors $\hat{\alpha}^c$ and $\hat{\beta}^c$ as $\hat{\beta}^c = \mu_c$ and $\hat{\alpha}^c = \upsilon_c \sigma_c$ ($c = 1,\ldots,p$), respectively.

■

Remark 5.1:
It is important to note that the algorithm intrinsically delivers estimates that satisfy the uniqueness condition $\|\hat{\beta}^c\|_2 = 1$, since the matrix μ_c in the SVD of $\hat{\Theta}_{\alpha\beta}$ is a unitary matrix.

■

Multi-channel Hammerstein model
Based on Algorithm 5.1, a multi-channel spatio-temporal Hammerstein model consisting of p channels

$$\hat{y}_n(x,t) = \sum_{c=1}^{p}\sum_{\tau=0}^{t}\int_{\Omega}\sum_{i=1}^{n}\sum_{j=1}^{m}\sum_{k=1}^{l}\hat{\alpha}^c_{i,j,k}\varphi_i(x)\psi_j(\zeta)\phi_k(\tau)\sum_{r=1}^{m}\sum_{s=1}^{v}\psi_r(\zeta)\hat{\beta}^c_s h_s(a_r(t-\tau))d\zeta\,, \quad (5.34)$$

is constructed to approximate the nonlinear DPS as shown in Figure 5.3. Each channel consists of the cascade connection of a static nonlinear block represented by basis functions, followed by a dynamic linear block represented by spatio-temporal Laguerre model as shown in Figure 5.4. The transfer functions in Figure 5.4 can be derived from (5.12) as follows

$$\kappa_1(s) = \frac{\sqrt{2\xi}}{s+\xi}, \kappa_2(s) = \cdots = \kappa_l(s) = \frac{s-\xi}{s+\xi}\,,$$

where ξ is the time-scaling factor.

Note that in Figure 5.3 and Figure 5.4, $\hat{y}_n(x,t) = \sum_{c=1}^{p}\hat{y}^c_n(x,t)$, $\hat{v}^c(\zeta,t) = \sum_{r=1}^{m}\sum_{s=1}^{v}\psi_r(\zeta)\hat{\beta}^c_s h_s(a_r(t))$, $\hat{D}^c_{j,k}(x) = \sum_{i=1}^{n}\hat{\alpha}^c_{i,j,k}\varphi_i(x)$ and $L^c_{j,k}(t) = \sum_{r=1}^{m}\sum_{s=1}^{v}\psi_{j,r}\hat{\beta}^c_s \ell_{k,s,r}(t)$.

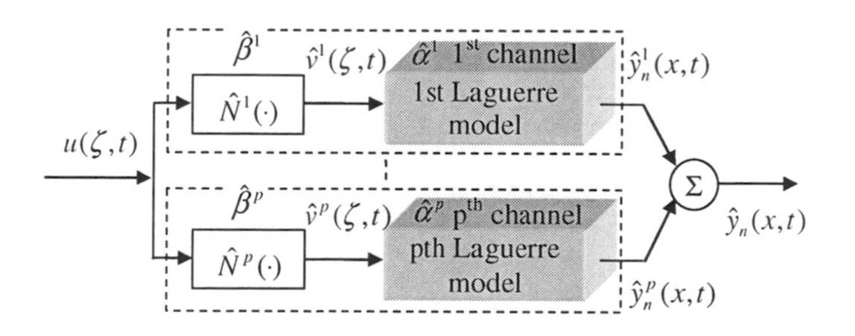

Fig. 5.3 Multi-channel spatio-temporal Hammerstein model

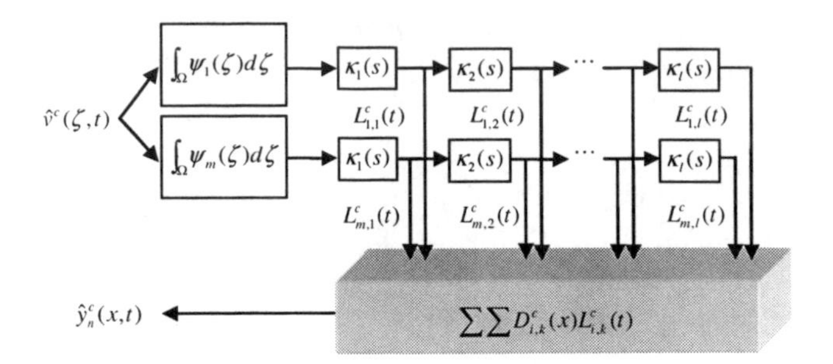

Fig. 5.4 Spatio-temporal Laguerre model of the cth channel

Remark 5.2:
If the time scale ξ is chosen suitably, then the Laguerre series can efficiently model any stable linear plant (Wang & Culett, 1995). Usually, the parameter ξ which gives a good performance is obtained from trials. Many studies have dealt with the time scale selection problem using such as offline optimization (Campello *et al.*, 2004) and online adaptation (Tanguy *et al.*, 2000) methods. For another parameter l, there are no theoretical methods but only some empirical ones so far. ■

Based on Lemma 5.1, we can give the following theorem to show the advantage of multi-channel mechanism.

Theorem 5.1:
For a spatio-temporal Hammerstein system (5.2), if the estimated matrix $\hat{\Theta}_{\alpha\beta}$ has rank $\gamma \geq 1$ and the parameters of the cth channel $(\hat{\beta}^c, \hat{\alpha}^c)$ $(c = 1,..., p, p \leq \gamma)$ are obtained by (5.31), then the parameter approximation error (5.32) will satisfy

$$\varepsilon^1 > \varepsilon^2 > \cdots \varepsilon^p > \cdots > \varepsilon^\gamma = 0 .$$

■

Proof: This can be easily drawn from Lemma 5.1.

■

Theorem 5.1 means that the parameter approximation error will be reduced by increasing the channel number p. Moreover, the model complexity can also be controlled by the number of channels. There is a tradeoff between the complexity and accuracy. Due to the property of the SVD, the parameter of the cth channel is the cth principal component of the whole parameter space. Therefore, only the first few dominant channels can construct a good model.

5.4.3 Convergence Analysis

An important issue is the convergence of the estimated parameters as the number of data points n_t tends to infinity. Now, we will give a convergence theorem to support the presented algorithm.

For simplicity, let $y_n(x,t) = H(x,t,\Theta,\{u(\zeta,\tau)\})$ denote a multi-channel Hammerstein model with n, l, $v < \infty$

$$y_n(x,t) = \sum_{c=1}^{p}\sum_{\tau=0}^{t}\int_{\Omega}\sum_{i=1}^{n}\sum_{j=1}^{m}\sum_{k=1}^{l}\alpha_{i,j,k}^{c}\varphi_i(x)\psi_j(\zeta)\phi_k(\tau)\sum_{r=1}^{m}\sum_{s=1}^{v}\psi_r(\zeta)\beta_s^c h_s(a_r(t-\tau))d\zeta\,, \quad (5.35)$$

where $\Theta = \sum_{c=1}^{p}[\alpha_{1,1}^{c}\beta_1^c,...,\alpha_{1,1}^{c}\beta_v^c,...,\alpha_{m,l}^{c}\beta_1^c,...,\alpha_{m,l}^{c}\beta_v^c]^T \in \mathbb{R}^{n\times mlv}$, $\alpha_{j,k}^{c}$ ($j=1,...,m$, $k=1,...,l$) are defined by (5.21), and $\{u(\zeta,\tau)\} = \{u(\zeta,\tau) \mid \zeta \in \Omega, \tau = 1,...,t\}$.

We always assume that there is an optimal model

$$y_n^*(x,t) = H(x,t,\Theta^*,\{u(\zeta,\tau)\})\,, \quad (5.36)$$

with an optimal parameter matrix Θ^* such that

$$\Theta^* = \arg\min_{\Theta\in D_\Theta}\{\bar{E}(y(x,t) - H(x,t,\Theta,\{u(\zeta,\tau)\}))^2\}\,, \quad (5.37)$$

where $\bar{E}f(x,t)^2 = \lim_{n_t\to\infty}\frac{1}{n_t}\sum_{t=1}^{n_t}\frac{1}{A}\int_{\Omega}Ef(x,t)^2dx$, $A = \int_{\Omega}dx$ and E is expectation operator. Let D_Θ be compact. Define $\Theta_{\alpha\beta}^* \in \mathbb{R}^{v\times nml}$ such that $\Theta^* = blockvec(\Theta_{\alpha\beta}^*)$.

Under the uniform spatial discretization, $\frac{1}{A}\sum_{i=1}^{\infty}b_i(t)^2 = \frac{1}{A}\int_{\Omega}y(x,t)^2dx$ can be replaced by $\frac{1}{n_y}\sum_{j=1}^{n_y}y(x_j,t)^2$. However, the accurate $a(t)$ can be obtained provided that $n_u \geq m$. Therefore, according to the details of the developed identification algorithm, the minimization problem (5.26) is indeed equivalent to the following problem

$$\hat{\Theta} = \arg\min_{\Theta\in D_\Theta}\{\frac{1}{n_t}\frac{1}{n_y}\sum_{t=1}^{n_t}\sum_{j=1}^{n_y}(y(x_j,t) - H(x_j,t,\Theta,\{u(\zeta,\tau)\}))^2\}\,. \quad (5.38)$$

It should be mentioned that, (5.26) can be considered as a practical implementation of (5.38) in order to reduce the involved spatial complexity. However, the theoretical analysis should be performed in the spatio-temporal domain.

Assumption 5.1:
Let $W(x,t)$ be the σ-algebra generated by ($d(x,t)$, $\cdots$, $d(x,0)$). For each t, τ ($t \geq \tau$) and any x, $\zeta \in \Omega$, there exist random variables $y_\tau^0(x,t)$ ($y_t^0(x,t) = 0$), $u_\tau^0(\zeta,t)$ ($u_t^0(\zeta,t) = 0$), that belong to $W(x,t)$, but are independent of $W(x,\tau)$, such that

$$E\mid y(x,t) - y_\tau^0(x,t)\mid^4 < M\lambda^{t-\tau}\,,$$
$$E\mid u(\zeta,t) - u_\tau^0(\zeta,t)\mid^4 < M\lambda^{t-\tau}\,,$$

for some $M < \infty$, $\lambda < 1$.

Assumption 5.2:
Assume that the model $y_n(x,t) = H(x,t,\Theta,\{u(\zeta,\tau)\})$ is differentiable with respect to Θ for all $\Theta \in D_\Theta$. Assume that

$$|H(x,t,\Theta,\{u_1(\zeta,\tau)\}) - H(x,t,\Theta,\{u_2(\zeta,\tau)\})| \leq M\sum_{\tau=0}^{t}\lambda^{t-\tau}\sup_{\zeta\in\Omega}|u_1(\zeta,\tau)-u_2(\zeta,\tau)|,$$

and $|H(x,t,\Theta,\{0(\zeta,\tau)\})| \leq M$, where Θ belongs to an open neighborhood of D_Θ, $M < \infty$ and $\lambda < 1$.

Assumption 5.3:
Define $\varepsilon(x,t,\Theta) = y(x,t) - H(x,t,\Theta,\{u(\zeta,\tau)\})$ and there exists

$$\left|\frac{\partial \varepsilon(x,t,\Theta)^2}{\partial \Theta}\right| \leq M\varepsilon(x,t,\Theta)^2,\ \Theta \in D_\Theta,\ \forall x \in \Omega,\ \forall t\,.$$

Remark 5.3:
Assumption 5.1 means that the system to be identified is exponentially stable, i.e., the remote past of the process is "forgotten" at an exponential rate. Assumption 5.2 has three meanings. First, the model is differential with respect to the parameters. Second, the model may not increase faster than the linear one. Third, the model is also exponentially stable. Regarding Assumption 5.3, the derivative of the modeling error with respect to the parameters is bounded by the modeling error. Such conditions are required to make the parameter optimization procedure feasible, and guarantee the following convergence.

Theorem 5.2:
For a spatio-temporal Hammerstein system (5.2), the multi-channel model (5.34) is estimated using Algorithm 5.1. If Assumption 5.1, Assumption 5.2 and Assumption 5.3 are satisfied, then $\sum_{c=1}^{p}\hat{\beta}^c(\hat{\alpha}^c)^T \to \Theta_{\alpha\beta}^*$ and $\hat{y}_n(x,t) \to y_n^*(x,t)$ w. p. 1 as $n_t \to \infty$, $n_y \to \infty$, and $p \to \gamma$, where $\gamma = rank(\Theta_{\alpha\beta}^*)$.

Proof:
In order to obtain the convergence with probability 1, the following lemma, which is the direct extension of the previous work (Cramér & Leadbetter, 1967; Ljung, 1978), is needed in the proof of Theorem 5.2.

Lemma 5.2:
Let $\xi(x,t)$ be a random variable with zero-mean value and with

$$|E(\xi(x,t)\xi(x,\tau))| \leq M\frac{t^\alpha + \tau^\beta}{1+|t-\tau|^\beta}, x \in \Omega, M < \infty, 0 \leq 2\alpha < \beta < 1\,. \tag{5.39}$$

Then

$$\frac{1}{n_t}\sum_{t=1}^{n_t}\xi(x,t)\to 0,\ \text{w. p. 1 as } n_t\to\infty\,. \tag{5.40}$$

where 'w. p. 1' means 'with probability 1'.

■

We now turn to the proof of Theorem 5.2. The convergence of the estimation $\hat{\Theta}$

$$\hat{\Theta}\to\Theta^*,\ \text{w. p. 1 as } n_t\to\infty,\ n_y\to\infty\,, \tag{5.41}$$

implies that

$$\hat{\Theta}_{\alpha\beta}\to\Theta^*_{\alpha\beta},\ \text{w. p. 1 as } n_t\to\infty,\ n_y\to\infty\,. \tag{5.42}$$

By Lemma 5.1, we have

$$\sum_{c=1}^{p}\hat{\beta}^c(\hat{\alpha}^c)^T\to\hat{\Theta}_{\alpha\beta},\ \text{as } p\to\gamma\,,$$

where $\gamma=rank(\hat{\Theta}_{\alpha\beta})$. Therefore

$$\sum_{c=1}^{p}\hat{\beta}^c(\hat{\alpha}^c)^T\to\Theta^*_{\alpha\beta},\ \text{w. p. 1 as } n_t\to\infty,\ n_y\to\infty,\ p\to\gamma\,. \tag{5.43}$$

Since $H(x,t,\Theta,\{u(\zeta,\tau)\})$ is continuous with respect to Θ , the convergence of parameters as in (5.43) naturally leads to the convergence of the model to its optimum

$$\hat{y}_n(x,t)\to y_n^*(x,t),\ \text{w. p. 1 as } n_t\to\infty,\ n_y\to\infty,\ p\to\gamma\,. \tag{5.44}$$

Define

$$Q_{n_yn_t}(\Theta)=\frac{1}{n_y}\sum_{j=1}^{n_y}\{\frac{1}{n_t}\sum_{t=1}^{n_t}\varepsilon(x_j,t,\Theta)^2\}\,.$$

As define in (5.37), Θ^* minimizes

$$\bar{E}(y(x,t)-H(x,t,\Theta,\{u(\zeta,\tau)\}))^2=\lim_{n_y\to\infty}\{\lim_{n_t\to\infty}EQ_{n_yn_t}(\Theta)\}\,,$$

and the estimate $\hat{\Theta}$ minimizes $Q_{n_yn_t}$ as defined in (5.38).

In order to prove (5.41), we should prove the convergence as follows

$$\sup_{\Theta\in D_\Theta}|Q_{n_yn_t}(\Theta)-EQ_{n_yn_t}(\Theta)|\to 0,\ \text{w. p. 1 as } n_t\to\infty,\ n_y\to\infty\,. \tag{5.45}$$

One feasible solution is to achieve the following convergence at any fixed spatial variable x before working at the spatio-temporal space.

$$\sup_{\Theta\in D_\Theta}\frac{1}{n_t}\sum_{t=1}^{n_t}|\varepsilon(x,t,\Theta)^2-E\varepsilon(x,t,\Theta)^2|\to 0,\ \text{w. p. 1 as } n_t\to\infty\,. \tag{5.46}$$

To achieve the convergence of (5.46), we have to obtain the convergence first at the pre-defined small open sphere, and then extend it to the global domain D_Θ using Heine-Borel's theorem.

Convergence of modeling error ε to its optimum over B :

Define the supremum between the model error and its optimum as a random variable

$$\eta(x,t)=\eta(x,t,\Theta^0,\rho)=\sup_{\Theta\in B}[\varepsilon(x,t,\Theta)^2-E\varepsilon(x,t,\Theta)^2].$$

Let D be the open neighborhood of D_Θ and choose $\Theta^0\in D_\Theta$. We can define a small open sphere centered at Θ^0 as

$$B(\Theta^0,\rho)=\{\Theta \,|\, |\Theta-\Theta^0|<\rho\}.$$

Let $B=B(\Theta^0,\rho)\cap D$, then

$$\sup_{\Theta\in B}\frac{1}{n_t}\sum_{t=1}^{n_t}[\varepsilon(x,t,\Theta)^2-E\varepsilon(x,t,\Theta)^2]\le\frac{1}{n_t}\sum_{t=1}^{n_t}\eta(x,t). \tag{5.47}$$

Define $\xi(x,t)=\eta(x,t)-E\eta(x,t)$. *If we can prove*

- $\xi(x,t)$ *satisfies Lemma 5.2 and*
- *the mean of $\eta(x,t)$ is infinitesimal,*

then $\eta(x,t)$ is also infinitesimal.

Firstly, we consider

$$|E(\xi(x,t)\xi(x,\tau))|=Cov[\eta(x,t),\eta(x,\tau)].$$

Define $\eta_\tau^0(x,t)=\sup_{\Theta\in B}[\varepsilon_\tau^0(x,t,\Theta)^2-E\varepsilon(x,t,\Theta)^2]$, with

$$\varepsilon_\tau^0(x,t,\Theta)=y_\tau^0(x,t)-H(x,t,\Theta,\{u_\tau^0(\zeta,j)\}),\ t>\tau,$$

where $\{u_\tau^0(\zeta,j)\}$ denotes the input set $(u_\tau^0(\zeta,t),...,u_\tau^0(\zeta,\tau+1),0,...,0)$ for all $\zeta\in\Omega$, $y_\tau^0(x,t)$ and $u_\tau^0(\zeta,j)$ are the variables introduced in Assumption 5.1. For convenience, let $u_\tau^0(\zeta,j)=0$ and $y_\tau^0(x,j)=0$ for $j<\tau$. Obviously $\eta_\tau^0(x,t)$ is independent of $\eta(x,\tau)$ from Assumption 5.1.

Hence

$$Cov[\eta(x,t),\eta(x,\tau)]=Cov[\eta(x,t)-\eta_\tau^0(x,t),\eta(x,\tau)].$$

Then using Schwarz's inequality, we have

$$|E(\xi(x,t)\xi(x,\tau))|\le[E\eta(x,\tau)^2E(\eta(x,t)-\eta_\tau^0(x,t))^2]^{1/2}. \tag{5.48}$$

Since

$$
\begin{aligned}
|\eta(x,t)-\eta_\tau^0(x,t)| &\le \sup_{\Theta\in B}|\varepsilon(x,t,\Theta)^2-\varepsilon_\tau^0(x,t,\Theta)^2| \\
&\le \sup_{\Theta\in B}\{|\varepsilon(x,t,\Theta)|+|\varepsilon_\tau^0(x,t,\Theta)|\}\times \sup_{\Theta\in B}|\varepsilon(x,t,\Theta)-\varepsilon_\tau^0(x,t,\Theta)|,
\end{aligned}
$$

using Assumption 5.2, we can further have

$$
\begin{aligned}
|\eta(x,t)-\eta_\tau^0(x,t)| \le M[&\sum_{j=0}^{t}\lambda^{t-j}\{|y(x,j)|+|y_\tau^0(x,j)|+\sup_{\zeta\in\Omega}|u(\zeta,j)|+\sup_{\zeta\in\Omega}|u_\tau^0(\zeta,j)|\}]\times \\
&[\sum_{j=0}^{t}\lambda^{t-j}\{|y(x,j)-y_\tau^0(x,j)|+\sup_{\zeta\in\Omega}|u(\zeta,j)-u_\tau^0(\zeta,j)|\}].
\end{aligned}
$$

Using Assumption 5.1 and Schwarz's inequality, we can finally derive

$$
E|\eta(x,t)-\eta_\tau^0(x,t)|^2 \le M\lambda^{t-\tau}. \tag{5.49}
$$

Following the similar derivation above and using Assumption 5.2 and Assumption 5.1, we can also derive

$$
E\eta(x,\tau)^2 \le M . \tag{5.50}
$$

Placing (5.49) and (5.50) into (5.48), we can easily derive that $\xi(x,t)$ satisfies Lemma 5.2, that is

$$
\frac{1}{n_t}\sum_{t=1}^{n_t}\xi(x,t)=\frac{1}{n_t}\sum_{t=1}^{n_t}(\eta(x,t)-E\eta(x,t))\to 0,\ \text{w. p. 1 as } n_t\to\infty . \tag{5.51}
$$

Secondly, we derive the mean value of η

$$
E\eta(x,t)=E\sup_{\Theta\in B}[\varepsilon(x,t,\Theta)^2-E\varepsilon(x,t,\Theta)^2].
$$

Since the right-hand side is continuous with respect to Θ, $E\eta(x,t)$ should be small if B is small. Furthermore, by Assumption 5.3,

$$
\left|\frac{\partial\varepsilon(x,t,\Theta)^2}{\partial\Theta}\right| \le M|\varepsilon(x,t,\Theta)|^2 \le M[\sum_{j=0}^{t}\lambda^{t-j}\{|y(x,j)|+\sup_{\zeta\in\Omega}|u(\zeta,j)|\}]^2,
$$

where we again have used the uniform bounds in Assumption 5.2. Consequently, by Assumption 5.1,

$$
E\sup_{\Theta\in B}\left|\frac{\partial\varepsilon(x,t,\Theta)^2}{\partial\Theta}\right|^2 \le M .
$$

Now

$$\begin{aligned} E\eta(x,t) &= E\sup_{\Theta\in B}[\varepsilon(x,t,\Theta)^2 - E\varepsilon(x,t,\Theta)^2] \\ &\le E\sup_{\Theta\in B}[\varepsilon(x,t,\Theta)^2 - \varepsilon(x,t,\Theta^0)^2] + \sup_{\Theta\in B} E[\varepsilon(x,t,\Theta^0)^2 - \varepsilon(x,t,\Theta)^2] \\ &\le [E\sup_{\Theta\in B}|\frac{\partial\varepsilon(x,t,\Theta)^2}{\partial\Theta}| + \sup_{\Theta\in B} E|\frac{\partial\varepsilon(x,t,\Theta)^2}{\partial\Theta}|]\times\sup_{\Theta\in B}|\Theta-\Theta^0| \\ &\le M^0\rho. \end{aligned} \tag{5.52}$$

Finally, from (5.52), (5.47) becomes

$$\sup_{\Theta\in B}\frac{1}{n_t}\sum_{t=1}^{n_t}[\varepsilon(x,t,\Theta)^2 - E\varepsilon(x,t,\Theta)^2] \le \frac{1}{n_t}\sum_{t=1}^{n_t}(\eta(x,t) - E\eta(x,t)) + M^0\rho. \tag{5.53}$$

It is clear to see from (5.51) that the first term of the right-hand side is arbitrarily small for sufficiently large n_t. Since ρ can also be arbitrarily small, therefore

$$\sup_{\Theta\in B}\frac{1}{n_t}\sum_{t=1}^{n_t}|\varepsilon(x,t,\Theta)^2 - E\varepsilon(x,t,\Theta)^2| \to 0,\ w.p.\ 1\ as\ n_t\to\infty. \tag{5.54}$$

Convergence extension to global D_Θ:

Since D_Θ is compact, by applying Heine-Borel's theorem, from (5.54) the following result is easily concluded

$$\sup_{\Theta\in D_\Theta}\frac{1}{n_t}\sum_{t=1}^{n_t}|\varepsilon(x,t,\Theta)^2 - E\varepsilon(x,t,\Theta)^2| \to 0,\ \text{w. p. 1 as } n_t\to\infty. \tag{5.55}$$

Extension to spatio-temporal domain:

Obviously

$$\sup_{\Theta\in D_\Theta}|Q_{n_y n_t}(\Theta) - EQ_{n_y n_t}(\Theta)| \le \frac{1}{n_y}\sum_{j=1}^{n_y}\sup_{\Theta\in D_\Theta}\frac{1}{n_t}\sum_{t=1}^{n_t}|\varepsilon(x_j,t,\Theta)^2 - E\varepsilon(x_j,t,\Theta)^2|,$$

therefore

$$\sup_{\Theta\in D_\Theta}|Q_{n_y n_t}(\Theta) - EQ_{n_y n_t}(\Theta)| \to 0,\ \text{w. p. 1 as } n_t\to\infty,\ n_y\to\infty. \tag{5.56}$$

■

5.5 Simulation and Experiment

In order to evaluate the presented modeling method, we first present the simulation on a typical distributed process, and then apply it to the snap curing oven.

For an easy comparison, some performance indexes are set up as follows:

- Spatio-temporal error $e(x,t) = y(x,t) - \hat{y}_n(x,t)$,
- Spatial normalized absolute error, $SNAE(t) = \int |e(x,t)| dx / \int dx$,
- Temporal normalized absolute error, $TNAE(x) = \sum |e(x,t)| / \sum \Delta t$,
- Root of mean squared error, $RMSE = (\int \sum e(x,t)^2 dx / \int dx \sum \Delta t)^{1/2}$.

5.5.1 Packed-Bed Reactor

Consider the packed-bed reactor given in Section 1.1.2. A dimensionless model that describes temperature distribution in the reactor is provided as follows (Christofides, 1998)

$$\begin{aligned} \varepsilon_p \frac{\partial y_g}{\partial t} &= -\frac{\partial y_g}{\partial x} + \alpha_c (y - y_g) - \alpha_g (y_g - \psi(x) a(t)), \\ \frac{\partial y}{\partial t} &= \frac{\partial^2 y}{\partial x^2} + \beta_0 e^{\frac{\gamma y}{1+y}} - \beta_c (y - y_g) - \beta_p (y - \psi(x) a(t)), \end{aligned} \tag{5.57}$$

subject to the boundary conditions

$$\begin{aligned} &x = 0,\ y_g = 0,\ \frac{\partial y}{\partial x} = 0, \\ &x = 1,\ \frac{\partial y}{\partial x} = 0, \end{aligned} \tag{5.58}$$

where y_g, y and a denote the dimensionless temperature of the gas, the catalyst and jacket, respectively. A small positive value ε_p denotes the ratio of the heat capacitance of the gas phase vs the heat capacitance of the catalytic phase. α_c, α_g, β_0, γ, β_c, and β_p are other system parameters. See Christofides (1998) for more details. It is assumed that only catalyst temperature measurements are available. The problem is to model the dynamics for the catalyst temperature.

The values of the process parameters are given below

$\varepsilon_p = 0.01$, $\gamma = 21.14$, $\beta_c = 1.0$, $\beta_p = 15.62$, $\beta_0 = -0.003$, $\alpha_c = 0.5$ and $\alpha_g = 0.5$.

A heater is used with the spatial distribution $\psi(x) = \sin(\pi x)$, $0 \le x \le 1$. In the numerical calculation, we set the input $a(t) = 1.1 + 1.5\sin(t/20)$. In the simulation, sixteen sensors are used in order to capture the sufficient spatial information. The random process noise is bounded by 0.003 with zero mean so that the noisy data for the modeling have a signal-to-noise ratio (SNR) of around 21dB. The sampling period Δt is 0.0001 and the simulation time is 0.13. A noisy data set of 1300 data is collected. The first 1100 data is used as the training data with the first 900 data as the estimation data and the next 200 data as the validation data. The validation data is used to monitor the training process and determine some design parameters using the cross-validation method. The remaining 200 data is the testing data.

The process output $y(x,t)$ is shown in Figure 5.5, while the obtained KL basis functions are shown in Figure 5.6 with $n=3$. Using the cross-validation the temporal bases $\phi_i(t)$, ($i=1,...,10$) are chosen as Laguerre series with time-scaling factor $\xi=20.5$. The nonlinear bases are polynomials as $h_i(a)=a^i$ ($i=1,...,4$).

The prediction output $\hat{y}_n(x,t)$ of the *3-channel Hammerstein model* is shown in Figure 5.7, with the prediction error $e(x,t)$ presented in Figure 5.8. It is obvious that the *3-channel Hammerstein model* can approximate the original spatio-temporal dynamics very well. As shown in Figure 5.9 and Figure 5.10, as the channel number increases, the prediction error $SNAE(t)$ and $TNAE(x)$ over the whole data set will decrease, which is consistent with the theoretical analysis. As illustrated in Figure 5.11, the *3-channel Hammerstein modeling* error $RMSE$ will become smaller when using more sensors. Numbers of sensors used should be determined by the complexity of spatial dynamics to be modeled.

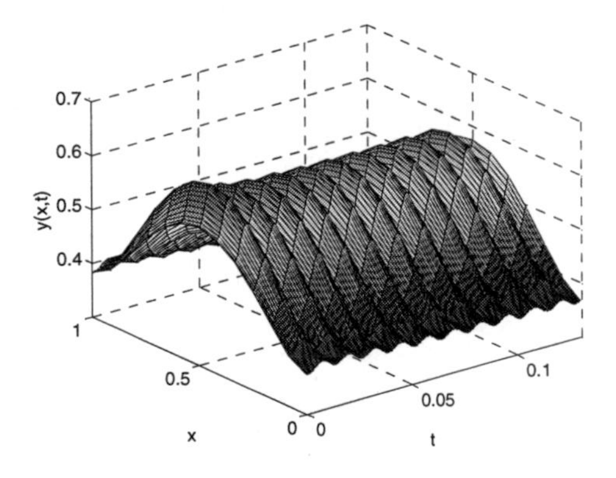

Fig. 5.5 Packed-bed reactor: Process output for multi-channel Hammerstein modeling

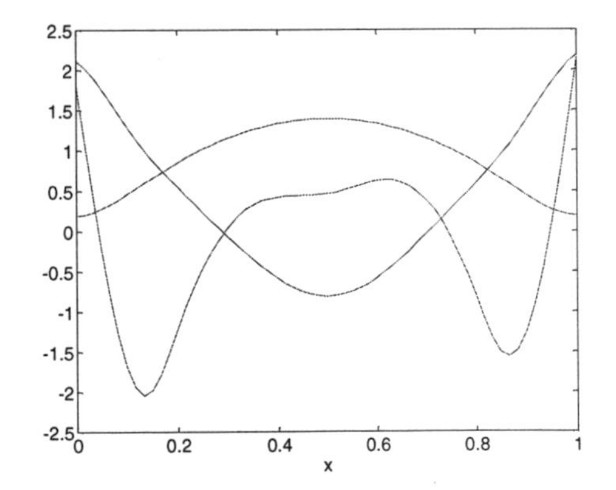

Fig. 5.6 Packed-bed reactor: KL basis functions for multi-channel Hammerstein modeling

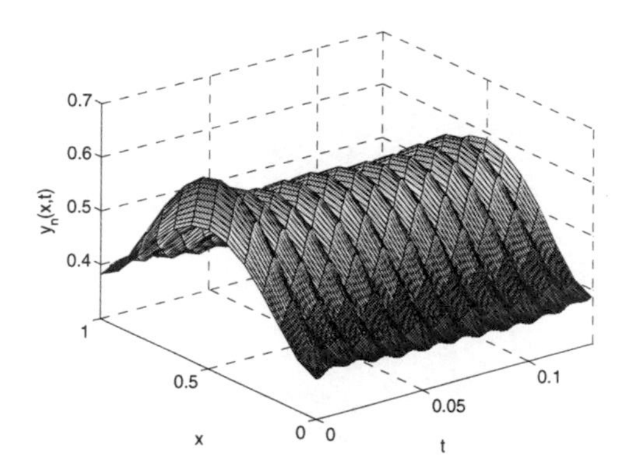

Fig. 5.7 Packed-bed reactor: Prediction output of 3-channel Hammerstein model

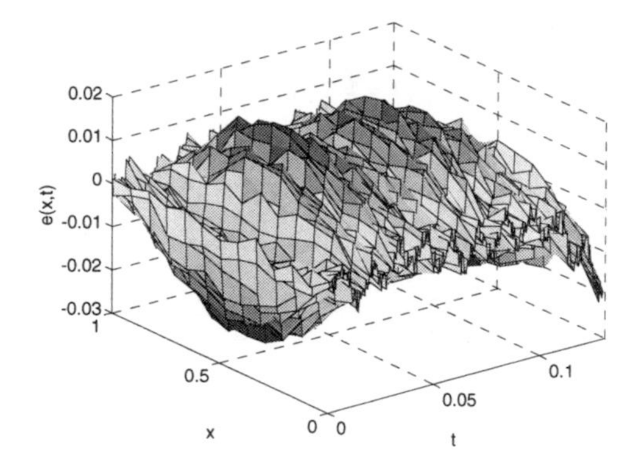

Fig. 5.8 Packed-bed reactor: Prediction error of 3-channel Hammerstein model

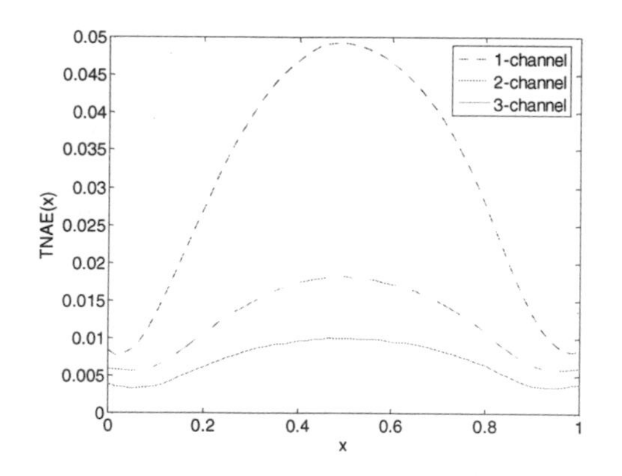

Fig. 5.9 Packed-bed reactor: *TNAE*(x) of Hammerstein models

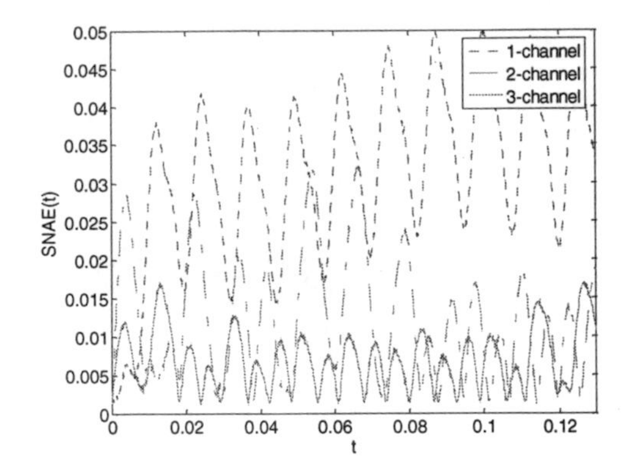

Fig. 5.10 Packed-bed reactor: *SNAE*(*t*) of Hammerstein models

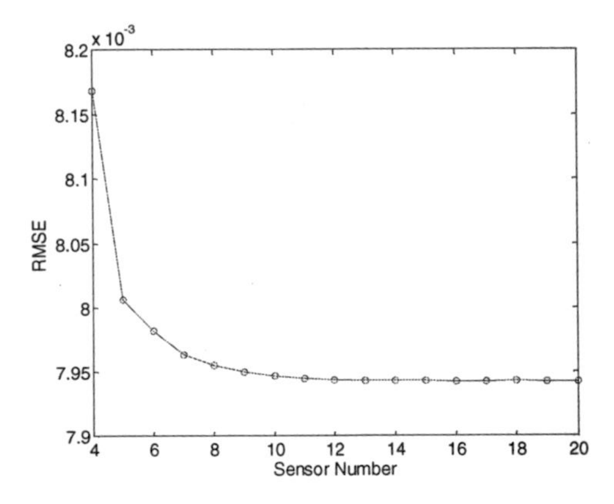

Fig. 5.11 Packed-bed reactor: *RMSE* of 3-channel Hammerstein model

5.5.2 Snap Curing Oven

Consider the snap curing oven (Figure 1.1 and Figure 3.11) provided in Sections 1.1.1 and 3.6.2. In the experiment, a total of 2400 measurements are collected with a sampling interval $\Delta t = 10$ seconds. One thousand and six hundred of measurements from sensors (s1-s5, s7-s10, and s12-s16) are used to estimate the model. The last 800 measurements from sensors (s1-s5, s7-s10, and s12-s16) are chosen to validate the model during the training. All 2400 measurements from the rest sensors (s6, s11) are used for modeling performance testing.

In the spatio-temporal Hammerstein modeling, five two-dimensional Karhunen-Loève basis functions are used as spatial bases and the first two of them are shown in Figure 5.12 and Figure 5.13. The temporal bases $\phi_i(t)$ are chosen as Laguerre series with the time-scaling factor $p = 0.001$ and the truncation length $q = 3$ using the cross-validation method.

The *3-channel Hammerstein model* is used to model the thermal process. After the training using the first 1600 data from the sensors (s1-s5, s7-s10, and s12-s16), a process model can be obtained with the significant performance such as the sensor s1 in Figure 5.14. The model also performs very well for the untrained locations such as the sensor s6 in Figure 5.15. The predicted temperature distribution of the oven at t=10000s is provided in Figure 5.16. The performance comparisons over the whole data set in Table 5.1 further show that the *3-channel Hammerstein model* has a much better performance than the *1- and 2-channel Hammerstein model.* The effectiveness of the presented modeling method is clearly demonstrated in this real application.

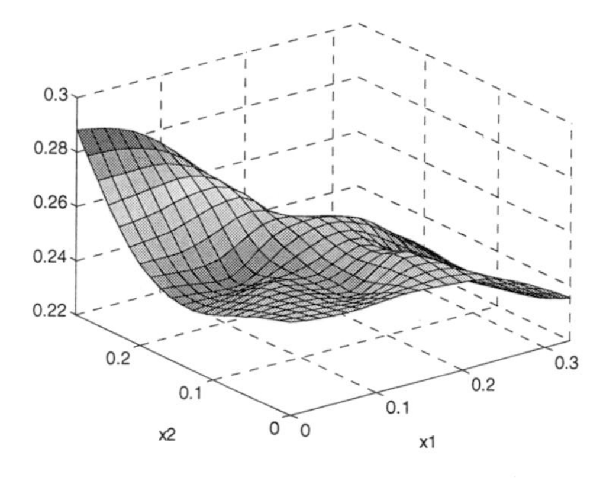

Fig. 5.12 Snap curing oven: KL basis functions (i=1) for multi-channel Hammerstein modeling

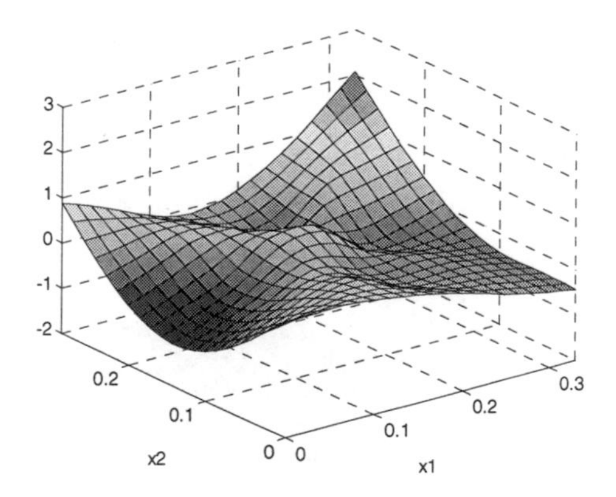

Fig. 5.13 Snap curing oven: KL basis functions (i=2) for multi-channel Hammerstein modeling

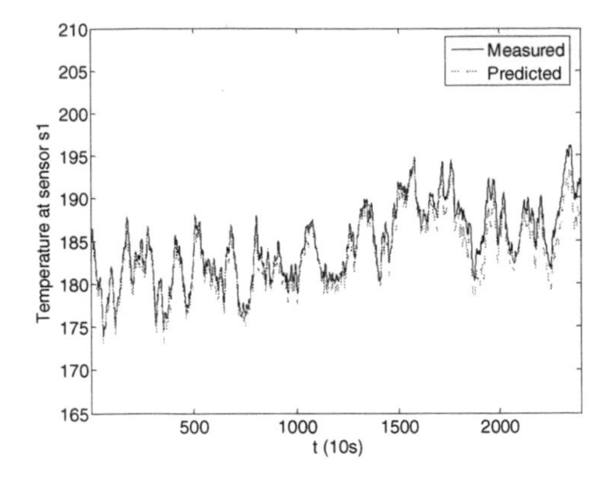

Fig. 5.14 Snap curing oven: Performance of 3-channel Hammerstein model at sensor s1

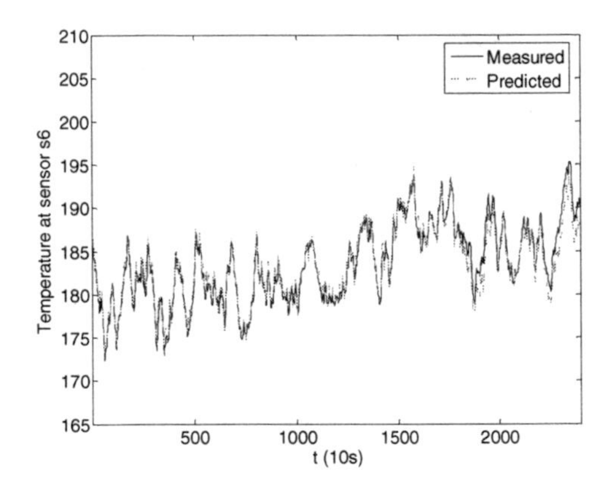

Fig. 5.15 Snap curing oven: Performance of 3-channel Hammerstein model at sensor s6

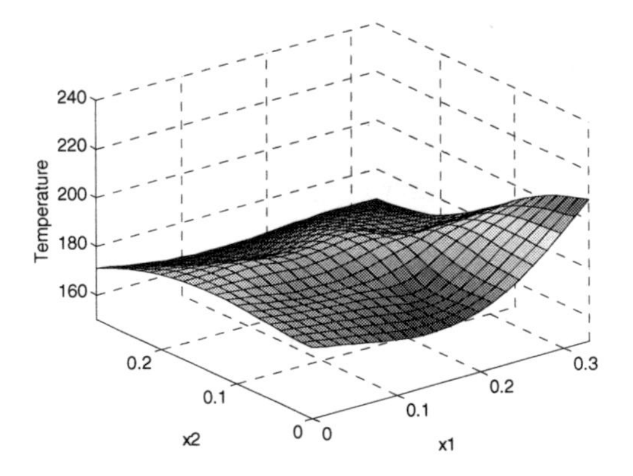

Fig. 5.16 Snap curing oven: Predicted temperature distribution of 3-channel Hammerstein model at t=10000s

Table 5.1 Snap curing oven: *TNAE*(x) of Hammerstein models

	s1	s2	s3	s4	s5	s6	s7	s8
1-channel model	2.24	1.91	2.31	1.74	2.6	1.87	1.96	1.8
2-channel model	1.39	1.26	1.64	1.21	2.03	0.94	1.26	1.63
3-channel model	1.17	1.05	1.32	0.86	1.91	0.73	0.95	1.37
	s9	s10	s11	s12	s13	s14	s15	s16
1-channel model	1.51	2.26	2.6	2.16	2.08	2.59	1.62	1.97
2-channel model	1.11	1.54	1.94	1.54	0.9	1.71	1.1	1.19
3-channel model	1.01	1.35	1.75	1.43	0.75	1.5	0.94	1.05

5.6 Summary

A novel multi-channel spatio-temporal Hammerstein modeling approach is presented for nonlinear distributed parameter systems. The Hammerstein distributed parameter model consists of the static nonlinear and the distributed dynamical linear parts. The distributed linear part is represented by a spatio-temporal kernel, i.e., Green's function. In the single-channel Hammerstein modeling, using the Galerkin method with the expansion onto KL spatial bases and Laguerre temporal bases, the spatio-temporal modeling is reduced to a traditional temporal modeling problem. The unknown parameters can be easily estimated using the least-squares estimation and the singular value decomposition. In the presence of unmodeled dynamics, a multi-channel Hammerstein modeling approach is used which can improve the modeling performance since the single-channel Hammerstein will lead to a relatively large modeling error. This modeling method provides convergent estimates under some conditions. The simulation example and the experiment on snap curing oven are presented to show the effectiveness of this modeling method and its potential to industrial applications.

References

[1] Bai, E.W.: An optimal two-stage identification algorithm for Hammerstein-Wiener nonlinear systems. Automatica 34(3), 333–338 (1998)

[2] Bai, E.W., Li, D.: Convergence of the iterative Hammerstein system identification algorithm. IEEE Transactions on Automatic Control 49(11), 1929–1940 (2004)

[3] Campello, R.J.G.B., Favier, G., Amaral, W.C.: Optimal expansions of discrete-time Volterra models using Laguerre functions. Automatica 40(5), 815–822 (2004)

[4] Chen, H.F.: Pathwise convergence of recursive identification algorithms for Hammerstein systems. IEEE Transactions on Automatic Control 49(10), 1641–1649 (2004)

[5] Christofides, P.D.: Robust control of parabolic PDE systems. Chemical Engineering Science 53(16), 2949–2965 (1998)

[6] Christofides, P.D.: Nonlinear and robust control of PDE systems: Methods and applications to transport-reaction processes. Birkhäuser, Boston (2001b)
[7] Cramér, H., Leadbetter, M.R.: Stationary and related stochastic processes. Wiely, New York (1967)
[8] Datta, K.B., Mohan, B.M.: Orthogonal functions in systems and control. World Scientific, Singapore (1995)
[9] Eskinat, E., Johnson, S., Luyben, W.: Use of Hammerstein models in identification of nonlinear systems. AIChE Journal 37(2), 255–268 (1991)
[10] Fruzzetti, K.P., Palazoglu, A., McDonald, K.A.: Nonlinear model predictive control using Hammerstein models. Journal of Process Control 7(1), 31–41 (1997)
[11] Golub, G., Van Loan, C.: Matrix computations, 3rd edn. Johns Hopkins University Press, Baltimore (1996)
[12] Gómez, J.C., Baeyens, E.: Identification of block-oriented nonlinear systems using orthonormal bases. Journal of Process Control 14(6), 685–697 (2004)
[13] Greblicki, W.: Continuous-time Hammerstein system identification from sampled data. IEEE Transactions on Automatic Control 51(7), 1195–1200 (2006)
[14] Heuberger, P.S.C., Van den Hof, P.M.J., Bosgra, O.H.: A generalized orthonormal basis for linear dynamical systems. IEEE Transactions on Automatic Control 40(3), 451–465 (1995)
[15] Ljung, L.: Convergence analysis of parametric identification methods. IEEE Transactions on Automatic Control 23(5), 770–783 (1978)
[16] Ljung, L.: System identification: Theory for the user, 2nd edn. Prentice-Hall, Englewood Cliffs (1999)
[17] Narendra, K., Gallman, P.: An iterative method for the identification of nonlinear systems using a Hammerstein model. IEEE Transactions on Automatic Control 11(3), 546–550 (1966)
[18] Park, H.M., Cho, D.H.: Low dimensional modeling of flow reactors. International Journal of Heat and Mass Transfer 39(16), 3311–3323 (1996a)
[19] Park, H.M., Cho, D.H.: The use of the Karhunen-Loève decomposition for the modeling of distributed parameter systems. Chemical Engineering Science 51(1), 81–98 (1996b)
[20] Sjöberg, J., Zhang, Q., Ljung, L., Benveniste, A., Delyon, B., Glorennec, P., Hjalmarsson, H., Juditsky, A.: Nonlinear black-box modeling in system identification: A unified approach. Automatica 31(12), 1691–1724 (1995)
[21] Stoica, P.: On the convergence of an iterative algorithm used for Hammerstein system identification. IEEE Transactions on Automatic Control 26(4), 967–969 (1981)
[22] Tanguy, N., Morvan, R., Vilbé, P., Calvez, L.C.: Online optimization of the time scale in adaptive Laguerre-based filters. IEEE Transactions on Signal Processing 48(4), 1184–1187 (2000)
[23] Wahlberg, B.: System identification using Laguerre models. IEEE Transactions on Automatic Control 36(5), 551–562 (1991)
[24] Wahlberg, B.: System identification using Kautz models. IEEE Transactions on Automatic Control 39(6), 1276–1282 (1994)
[25] Wang, L., Cluett, W.R.: Optimal choice of time-scaling factor for linear system approximation using Laguerre models. IEEE Transactions on Automatic Control 39(7), 1463–1467 (1995)

[26] Vörös, J.: Recursive identification of Hammerstein systems with discontinuous nonlinearities containing dead-zones. IEEE Transactions on Automatic Control 48(12), 2203–2206 (2003)
[27] Zervos, C.C., Dumont, G.A.: Deterministic adaptive control based on Laguerre series representation. International Journal of Control 48(6), 2333–2359 (1988)
[28] Zhu, Y.C.: Identification of Hammerstein models for control using ASYM. International Journal of Control 73(18), 1692–1702 (2000)

6 Spatio-Temporal Volterra Modeling for a Class of Nonlinear DPS

Abstract. To model the nonlinear distributed parameter system (DPS), a spatio-temporal Volterra model is presented with a series of spatio-temporal kernels. It can be considered as a nonlinear generalization of Green's function or a spatial extension of the traditional Volterra model. To obtain a low-order model, the Karhunen-Loève (KL) method is used for the time/space separation and dimension reduction. Then the model can be estimated with a least-squares algorithm with the convergence guaranteed under noisy measurements. The simulation and experiment are conducted to demonstrate the effectiveness of the presented modeling method.

6.1 Introduction

In general, the linear distributed parameter system (DPS) can be represented using the impulse response function (i.e., Green's function and kernel). In some cases, the Green's function can be derived from the first-principle knowledge (Butkovskiy, 1982). On the other hand, when the analytical Green's function is not available, it can be estimated from the input-output data (Gay & Ray, 1995; Zheng, Hoo & Piovoso, 2002; Zheng & Hoo, 2002, 2004; Doumanidis & Fourligkas, 2001). However, the Green's function model uses one single kernel, which can only approximate the nonlinear system around the given working condition.

In the modeling of traditional lumped parameter systems (LPS), the fading memory nonlinear system (FMNS) (Boyd & Chua, 1985) has been proposed to cover a wide range of industrial processes, whose dependence on past inputs decreases rapidly with time in large-scale range around the working point. The dynamics of the FMNS can be modeled by the Volterra series to any desired accuracy (Boyd & Chua, 1985) because Volterra series are constructed by a series of kernels, from the 1st-order, the 2nd-order, and up to the high-order kernels. Extensive research has been reported for lumped system identification and control design using the Volterra model (Schetzen, 1980; Rugh, 1981; Doyle III *et al.*, 1995; Maner *et al.*, 1996; Parker *et al.*, 2001). However, the Volterra model is only studied in LPS, because the traditional Volterra series do not have spatio-temporal nature.

Similarly, many distributed parameter processes may have the features of FMNS. In order to model unknown nonlinear distributed parameter systems, the spatio-temporal kernel-based idea from the Green's function will be expanded into the Volterra series. A spatio-temporal Volterra model can be constructed by adding the space variables into the traditional Volterra model. Thus, this kind of

H.-X. Li and C. Qi: Spatio-Temporal Modeling of Nonlinear DPS, ISCA 50, pp. 123–147.
springerlink.com

spatio-temporal Volterra model should be capable to model a wide range of nonlinear DPS with stable dynamics and fading memory features. Since no feedback is involved, the Volterra model is guaranteed to be stable.

The spatio-temporal Volterra modeling approach is designed as follows. Firstly, the unknown nonlinear DPS to be estimated is expressed by a spatio-temporal Volterra model with a set of spatio-temporal kernels. In order to estimate the spatio-temporal kernels from the input-output data, each kernel is expanded onto spatial and temporal basis functions with unknown coefficients. To reduce the parametric complexity, the Karhunen-Loève (KL) method is used to find the dominant spatial basis functions and the Laguerre polynomials are selected as the temporal basis functions. Secondly, using the Galerkin method, this spatio-temporal modeling problem will turn into a temporal modeling problem. Thirdly, unknown parameters can be easily estimated using the least-squares method in the time domain. After the time/space synthesis of kernels, the spatio-temporal Volterra model can be constructed. Moreover, the state space representation of spatio-temporal Volterra model can be easily obtained. The convergent estimation can be guaranteed under certain conditions. The simulation and the experiment demonstrate the effectiveness of the presented modeling method.

This chapter is organized as follows. The spatio-temporal Volterra model is presented in Section 6.2. Section 6.3 presents the spatio-temporal Volterra modeling approach. The state space realization is provided in Section 6.4. Section 6.5 gives the convergence analysis. The simulation and the experiment are demonstrated in Section 6.6.

6.2 Spatio-Temporal Volterra Model

It is well known that a linear continuous DPS can be represented as a linear mapping from the input $u(x,t)$ to the output $y(x,t)$, where $x \in \Omega$ denotes space variable, Ω is the spatial domain, and t is time. This mapping can be expressed in a Fredholm integral equation of the first kind containing a square-integrable kernel g (i.e., impulse response function or Green's function) (Butkovskiy, 1982; Gay & Ray, 1995)

$$y(x,t) = \int_{\Omega} \int_{0}^{t} g(x,\zeta,t,\tau)u(\zeta,\tau)d\tau d\zeta \,. \tag{6.1}$$

On the other hand, a lumped parameter system

$$y(t) = N(\{u(\tau)\}) + d(t) \,,$$

where $\{u(\tau)\} = \{u(\tau) \mid \tau = 1,...,t\}$ is the input, t is the discretized time instant, y and d are the output and the stochastic disturbance, and N is an operator with the fading memory feature, can be approximated by a discrete-time Volterra model (Boyd & Chua, 1985)

$$y(t) = \sum_{r=1}^{\infty} \sum_{\tau_1=0}^{t} \cdots \sum_{\tau_r=0}^{t} g_r(t,\tau_1,...,\tau_r) \prod_{v=1}^{r} u(\tau_v) \,, \tag{6.2}$$

where g_r is the rth-order temporal kernel.

Motivated by (6.1) and (6.2), for a distributed parameter system

$$y(x,t) = N(\{u(\zeta,\tau)\}) + d(x,t)\,, \tag{6.3}$$

where $\{u(\zeta,\tau)\} = \{u(\zeta,\tau) \mid \zeta \in \Omega, \tau = 1,...,t\}$ is the input, a spatio-temporal Volterra model is constructed by adding the space variables into the traditional Volterra model

$$y(x,t) = \sum_{r=1}^{\infty} \int_{\Omega} \cdots \int_{\Omega} \sum_{\tau_1=0}^{t} \cdots \sum_{\tau_r=0}^{t} g_r(x,\zeta_1,...,\zeta_r,t,\tau_1,...,\tau_r) \prod_{v=1}^{r} u(\zeta_v,\tau_v) d\zeta_v\,, \tag{6.4}$$

where g_r is the rth-order spatio-temporal Volterra kernel, which denotes the influence of the input at the location $\zeta_1,...,\zeta_r$ and the time $\tau_1,...,\tau_r$ on the output at the location x and the time t. For easy understanding, the integral operator is used for the spatial operation and the sum operator for the temporal operation. Similar to (6.1), it is reasonable that the time/space variables in (6.4) are symmetrical. Obviously, the Green's function model is a first-order spatio-temporal Volterra model, and the kernels of the spatio-temporal Volterra model can be seen as the high-dimensional generalizations of the Green's function. Actually, the form of spatio-temporal Volterra model (6.4) can be derived for the DPS using Taylor expansion as traditional Volterra series derivation (Rugh, 1981), which is not included here for simplicity.

The model (6.4) can work for both the time-varying and the time-invariant systems. For the time-invariant system, the kernel will be invariant and represented as follows

$$g_r(x,\zeta_1,...,\zeta_r,t,\tau_1,...,\tau_r) = g_r(x,\zeta_1,...,\zeta_r,t-\tau_1,...,t-\tau_r)\,. \tag{6.5}$$

Similarly the model (6.4) can also work for the space-varying or space-invariant system. When the model is homogeneous in the space domain, there exists

$$g_r(x,\zeta_1,...,\zeta_r,t,\tau_1,...,\tau_r) = g_r(x-\zeta_1,...,x-\zeta_r,t,\tau_1,...,\tau_r)\,. \tag{6.6}$$

In this study, we only consider the time-invariant and space-varying case (6.5) since it is very common in the real applications. Substituting (6.5) into (6.4) will have the following expression

$$y(x,t) = \sum_{r=1}^{\infty} \int_{\Omega} \cdots \int_{\Omega} \sum_{\tau_1=0}^{t} \cdots \sum_{\tau_r=0}^{t} g_r(x,\zeta_1,...,\zeta_r,\tau_1,...,\tau_r) \prod_{v=1}^{r} u(\zeta_v,t-\tau_v) d\zeta_v\,. \tag{6.7}$$

The model (6.7) is still not applicable because of its infinite-order. In practice, the higher order terms can be neglected and only the first R kernels need to be taken into account as below

$$y(x,t) = \sum_{r=1}^{R} \int_{\Omega} \cdots \int_{\Omega} \sum_{\tau_1=0}^{t} \cdots \sum_{\tau_r=0}^{t} g_r(x,\zeta_1,...,\zeta_r,\tau_1,...,\tau_r) \prod_{v=1}^{r} u(\zeta_v,t-\tau_v) d\zeta_v + \upsilon(x,t), \tag{6.8}$$

where the error term $\upsilon(x,t)$ includes unmodeled dynamics and external noise. The modeling accuracy and the model complexity can be controlled by the order R.

6.3 Spatio-Temporal Modeling Approach

Now the problem is to estimate a spatio-temporal Volterra model (6.8) from a set of spatio-temporal input-output data $\{u(x,t) \mid x \in \Omega, t = 1,...,n_t\}$, $\{y(x_j,t) \mid x_j \in \Omega, j = 1,...,n_y, \ t = 1,...,n_t\}$, where n_t denotes the time length and n_y is the number of sampled spatial points of the output. For simplicity, it is assumed that the spatial information of the input is known from some physical knowledge and the locations x_j ($j = 1,...,n_y$) are uniformly distributed over the spatial domain. In order to achieve a good modeling performance, the order R can be determined in an incremental way using the cross-validation technique or some optimization methods. Once the order is determined, the next problem is to estimate the kernels. The main difficulty comes from the time/space coupling of kernels.

Using a simple time/space discretization for kernels $g_r(x,\zeta_1,...,\zeta_r,\tau_1,...,\tau_r)$, ($r = 1,...,R$) will lead to a large amount of parameters to be estimated. However, it is important to reduce the parametric complexity, improve the numerical condition and decrease the variance of the estimated parameters. This can be done using the time/space method, i.e., expanding the kernels in terms of a relatively small number of orthonormal basis functions such as KL spatial bases and Laguerre temporal bases. After the time/space separation, the original spatio-temporal problem will turn to the traditional temporal modeling problem. Thus, the unknown parameters can be easily estimated in the temporal domain. Finally, the spatio-temporal Volterra model can be reconstructed using the time/space synthesis. The modeling idea is shown in Figure 6.1. The time/space separation and synthesis are very critical for this identification approach, which are the key differences from the traditional Volterra modeling.

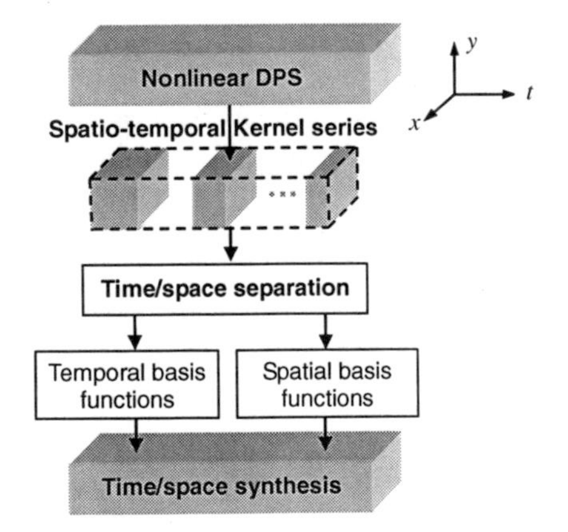

Fig. 6.1 Spatio-temporal Volterra modeling approach

6.3.1 Time/Space Separation

For simplicity, the input $u(x,t)$ is assumed to have a finite-dimensional freedom since only a finite number of actuators are available in practice. Therefore the input $u(x,t)$ can be formulated in terms of a finite number of spatial input basis functions $\{\psi_i(x)\}_{i=1}^m$

$$u(x,t)=\sum_{i=1}^{m}\psi_i(x)a_i(t)\,, \tag{6.9}$$

where $a_i(t)=\int_\Omega u(x,t)\psi_i(x)dx$ is the time coefficient (input signal) and m is the number of actuators. Ideally, the output $y(x,t)$ and the error $\upsilon(x,t)$ can be expressed by an infinite set of orthonormal spatial output basis functions $\{\varphi_i(x)\}_{i=1}^\infty$

$$y(x,t)=\sum_{i=1}^{\infty}\varphi_i(x)b_i(t)\,, \tag{6.10}$$

$$\upsilon(x,t)=\sum_{i=1}^{\infty}\varphi_i(x)\upsilon_i(t)\,, \tag{6.11}$$

where $b_i(t)=\int_\Omega y(x,t)\varphi_i(x)dx$ and $\upsilon_i(t)=\int_\Omega \upsilon(x,t)\varphi_i(x)dx$ are the time coefficients of the output and error respectively. This is because of the inherently infinite-dimensional characteristic of the DPS. Practically, for most of parabolic systems, both output $y(x,t)$ and error $\upsilon(x,t)$ can be truncated into n dimensions as below

$$y_n(x,t)=\sum_{i=1}^{n}\varphi_i(x)b_i(t)\,, \tag{6.12}$$

$$\upsilon_n(x,t)=\sum_{i=1}^{n}\varphi_i(x)\upsilon_i(t)\,. \tag{6.13}$$

The dimension n will be dependent on how the eigenspectrum of the DPS is separated into slow and fast modes, the type of spatial basis functions and the required modeling accuracy. For convenience, define $\boldsymbol{a}(t)=[a_1(t),\cdots,a_m(t)]^T\in\mathbb{R}^m$, $\boldsymbol{b}(t)=[b_1(t),\cdots,b_n(t)]^T\in\mathbb{R}^n$ and $\boldsymbol{\upsilon}(t)=[\upsilon_1(t),\cdots,\upsilon_n(t)]^T\in\mathbb{R}^n$.

Assuming that the kernels in (6.8) are absolutely integrable on the time domain $[0,\infty)$ at any spatial point x and ζ, which means that the corresponding spatio-temporal Volterra model is stable, then they can be represented by means of orthonormal temporal basis functions. The kernels are supposed to be expanded onto input bases $\{\psi_i(x)\}_{i=1}^m$, output bases $\{\varphi_i(x)\}_{i=1}^n$ and temporal bases $\{\phi_i(t)\}_{i=1}^q$

$$g_r(\cdot)=\sum_{i=1}^{n}\sum_{j_1=1}^{m}\cdots\sum_{j_r=1}^{m}\sum_{k_1=1}^{q}\cdots\sum_{k_r=1}^{q}\theta_{ij_1\ldots j_rk_1\ldots k_r}^{(r)}(\varphi_i(x)\prod_{v=1}^{r}\psi_{j_v}(\zeta_v)\phi_{k_v}(\tau_v))\,, \tag{6.14}$$

where $\theta^{(r)}_{ij_1\cdots j_r k_1\cdots k_r}$ is the corresponding constant coefficient of the rth-order kernel onto output bases $\varphi_i(x)$, input bases $\psi_{j_1}(\zeta_1),\cdots,\psi_{j_r}(\zeta_r)$ and temporal bases $\phi_{k_1}(\tau_1),\cdots,\phi_{k_r}(\tau_r)$, n is the dimension of output bases and q is the dimension of temporal bases. Theoretically, both n and q should be infinite for the DPS. Practically, for most of parabolic systems, finite n and q are often enough for a realistic approximation. Obviously they are affected by the required modeling accuracy. On the other hand, n is also dependent on the slow/fast eigenspectrum of the DPS and the type of spatial basis functions; and q is also related to the complexity of the system dynamics.

Selection of basis functions

The choice of basis functions will have a significant effect on the modeling performance, and the selections are summarized as follows.

(1) $\varphi_i(x)$ is usually selected as standard orthonormal functions such as Fourier series, Legendre polynomials, Jacobi polynomials and Chebyshev polynomials (Datta & Mohan, 1995). In this study, the KL method (Park & Cho, 1996a, 1996b) is chosen to identify the empirical spatial basis functions from the representative process data because fewer parameters need to be estimated in the Volterra modeling. Usually, the "energy method" is used to determine the small value of n. See Section 3.4 for more details.

(2) $\psi_i(x)$ is often determined from some physical knowledge, which describes the distribution of the control action $a_i(t)$ in the spatial domain Ω.

(3) $\phi_i(t)$ is often chosen as Laguerre function, Kautz function (Wahlberg, 1991; Wahlberg, 1994; Wahlberg & Mäkilä, 1996) or generalized orthonormal basis function (Heuberger, Van den Hof, & Bosgra, 1995). Here, Laguerre function is chosen for the development, because of its simplicity, and robustness to the sampling interval and the model dimension (Wahlberg, 1991). Laguerre functions are defined as a functional series (Zervos & Dumont, 1988)

$$\phi_i(t) \triangleq \sqrt{2p}\frac{e^{pt}}{(i-1)!}\cdot\frac{d^{i-1}}{dt^{i-1}}[t^{i-1}\cdot e^{-2pt}],\ i=1,2,\ldots,\infty,\ p>0, \tag{6.15}$$

where p is the time-scaling factor, and $t\in[0,\infty)$ is a time variable. The Laplace transform of the ith Laguerre function is given by

$$\phi_i(s)=\sqrt{2p}\frac{(s-p)^{i-1}}{(s+p)^i},\ i=1,2,\ldots,\infty,\ p>0. \tag{6.16}$$

Laguerre functions (6.15) and (6.16) form a complete orthonormal basis in the function space $L_2(0,\infty)$ and $H_2(C_+)$ respectively.

6.3.2 Temporal Modeling Problem

Substitution of (6.9) and (6.12)-(6.14) into (6.8) will have

$$\sum_{h=1}^{n}\varphi_h(x)b_h(t)=\sum_{r=1}^{R}\int_\Omega\cdots\int_\Omega\sum_{\tau_1=0}^{t}\cdots\sum_{\tau_r=0}^{t}\sum_{i=1}^{n}\sum_{j_1=1}^{m}\cdots\sum_{j_r=1}^{m}\sum_{k_1=1}^{q}\cdots\sum_{k_r=1}^{q}\theta^{(r)}_{ij_1\ldots j_rk_1\ldots k_r} \times\varphi_i(x)\prod_{v=1}^{r}\psi_{j_v}(\zeta_v)\phi_{k_v}(\tau_v)\sum_{w_v=1}^{m}\psi_{w_v}(\zeta_v)a_{w_v}(t-\tau_v)d\zeta_v+\sum_{h=1}^{n}\varphi_h(x)\upsilon_h(t). \tag{6.17}$$

Equation (6.17) can be further simplified into

$$\sum_{h=1}^{n}\varphi_h(x)b_h(t)=\sum_{r=1}^{R}\sum_{i=1}^{n}\varphi_i(x)\sum_{j_1=1}^{m}\cdots\sum_{j_r=1}^{m}\sum_{k_1=1}^{q}\cdots\sum_{k_r=1}^{q}\theta^{(r)}_{ij_1\ldots j_rk_1\ldots k_r}\prod_{v=1}^{r}\sum_{w_v=1}^{m}\psi_{j_vw_v}l_{k_vw_v}(t) +\sum_{h=1}^{n}\varphi_h(x)\upsilon_h(t). \tag{6.18}$$

where

$$\psi_{jw}=\int_\Omega\psi_j(\zeta)\psi_w(\zeta)d\zeta\,, \tag{6.19}$$

$$l_{kw}(t)=\sum_{\tau=0}^{t}\phi_k(\tau)a_w(t-\tau)\,. \tag{6.20}$$

Using the Galerkin method (Christofides, 2001b), projecting (6.18) onto the output basis functions $\varphi_{h_1}(x)$ ($h_1=1,\ldots,n$) will lead to the following expression

$$\int_\Omega\varphi_{h_1}(x)\sum_{h_2=1}^{n}\varphi_{h_2}(x)b_{h_2}(t)dx=\sum_{r=1}^{R}\sum_{i=1}^{n}\int_\Omega\varphi_{h_1}(x)\varphi_i(x)dx\sum_{j_1=1}^{m}\cdots\sum_{j_r=1}^{m}\sum_{k_1=1}^{q}\cdots\sum_{k_r=1}^{q}\theta^{(r)}_{ij_1\ldots j_rk_1\ldots k_r} \times\prod_{v=1}^{r}\sum_{w_v=1}^{m}\psi_{j_vw_v}l_{k_vw_v}(t)+\sum_{i=1}^{n}\int_\Omega\varphi_{h_1}(x)\varphi_i(x)dx\upsilon_i(t).$$

Re-arranging the order of integration and summation, it becomes

$$\sum_{h_2=1}^{n}\varphi_{h_1h_2}b_{h_2}(t)=\sum_{r=1}^{R}\sum_{i=1}^{n}\varphi_{h_1i}\sum_{j_1=1}^{m}\cdots\sum_{j_r=1}^{m}\sum_{k_1=1}^{q}\cdots\sum_{k_r=1}^{q}\theta^{(r)}_{ij_1\ldots j_rk_1\ldots k_r}\prod_{v=1}^{r}\sum_{w_v=1}^{m}\psi_{j_vw_v}l_{k_vw_v}(t) +\sum_{i=1}^{n}\varphi_{h_1i}\upsilon_i(t). \tag{6.21}$$

where $\varphi_{h_1h_2}=\int_\Omega\varphi_{h_1}(x)\varphi_{h_2}(x)dx$.

Since the matrix $\{\varphi_{h_1h_2}\}$ is invertible due to the orthonormal bases, the following expression can be derived from (6.21).

$$\boldsymbol{b}(t)=\sum_{r=1}^{R}\sum_{j_1=1}^{m}\cdots\sum_{j_r=1}^{m}\sum_{k_1=1}^{q}\cdots\sum_{k_r=1}^{q}\boldsymbol{\theta}^{(r)}_{j_1\ldots j_rk_1\ldots k_r}\prod_{v=1}^{r}\sum_{w_v=1}^{m}\psi_{j_vw_v}l_{k_vw_v}(t)+\boldsymbol{\upsilon}(t)\,. \tag{6.22}$$

where

$$\boldsymbol{\theta}_{j_1\ldots j_r k_1\ldots k_r}^{(r)} = [\theta_{1j_1\ldots j_r k_1\ldots k_r}^{(r)}, \cdots, \theta_{nj_1\ldots j_r k_1\ldots k_r}^{(r)}]^T \in \mathbb{R}^n . \tag{6.23}$$

6.3.3 Parameter Estimation

Equation (6.22) can be expressed in a linear regression form

$$\boldsymbol{b}(t) = \boldsymbol{\Theta}^T \boldsymbol{\Phi}(t) + \boldsymbol{\upsilon}(t) , \tag{6.24}$$

where

$$\boldsymbol{\Theta} = [\boldsymbol{\Theta}^{(1)}, \boldsymbol{\Theta}^{(2)}, \cdots, \boldsymbol{\Theta}^{(R)}]^T \in \mathbb{R}^{(mq+\cdots+m^R q^R)\times n} , \tag{6.25}$$

$$\boldsymbol{\Phi} = [\boldsymbol{\Phi}^{(1)}, \boldsymbol{\Phi}^{(2)}, \cdots, \boldsymbol{\Phi}^{(R)}]^T \in \mathbb{R}^{mq+\cdots+m^R q^R} , \tag{6.26}$$

and

$$\boldsymbol{\Theta}^{(r)} = [\boldsymbol{\theta}_{1\ldots11\ldots1}^{(r)} \cdots \boldsymbol{\theta}_{m\ldots mq\ldots q}^{(r)}] \in \mathbb{R}^{n\times m^r q^r} ,$$

$$\boldsymbol{\Phi}^{(r)} = [l_{1\ldots11\ldots1}^{(r)} \cdots l_{m\ldots mq\ldots q}^{(r)}] \in \mathbb{R}^{m^r q^r} ,$$

$$l_{j_1\ldots j_r k_1\ldots k_r}^{(r)}(t) = \prod_{v=1}^{r} \sum_{w_v=1}^{m} \psi_{j_v w_v} l_{k_v w_v}(t) \in \mathbb{R} .$$

In practice, u and y are uniformly sampled over the spatial domain. In this case, $\boldsymbol{a}$ and $\boldsymbol{b}$ can also be computed from the pointwise data using spline interpolation in the spatial domain. Then, $\boldsymbol{\Phi}(t)$ can be constructed from $\boldsymbol{a}$.

Now considering the n_t set of temporal data $\{\boldsymbol{a}(t) \mid t = 1,\ldots,n_t\}$, $\{\boldsymbol{b}(t) \mid t = 1,\ldots,n_t\}$, it is well known from (Ljung, 1999) that by minimizing a quadratic criterion of the prediction errors

$$\hat{\boldsymbol{\Theta}} = \arg\min_{\Theta}\{\frac{1}{n_t}\sum_{t=1}^{n_t} \| \boldsymbol{b}(t) - \boldsymbol{\Theta}^T \boldsymbol{\Phi}(t) \|^2\} , \tag{6.27}$$

where $\| f(t) \|^2 = f^T(t) f(t)$, $\boldsymbol{\Theta}$ can be estimated using the least-squares method as follows

$$\hat{\boldsymbol{\Theta}} = (\frac{1}{n_t}\sum_{t=1}^{n_t} \boldsymbol{\Phi}(t)\boldsymbol{\Phi}^T(t))^{-1}(\frac{1}{n_t}\sum_{t=1}^{n_t} \boldsymbol{\Phi}(t)\boldsymbol{b}^T(t)) , \tag{6.28}$$

provided that the indicated inverse exists. This condition can be guaranteed when using persistently exciting input.

After the kernels in (6.14) are reconstructed using the time/space synthesis from the estimated parameters $\hat{\boldsymbol{\Theta}}$, the spatio-temporal Volterra model can be obtained from (6.8).

6.4 State Space Realization

The spatio-temporal Volterra model (6.8) can also be transformed into a state space form. The Laguerre network representation of the realization procedure is shown in Figure 6.2, where the transfer functions

$$G_1(s)=\frac{\sqrt{2p}}{s+p}, G_2(s)=\cdots=G_q(s)=\frac{s-p}{s+p},$$

can be derived from (6.16), and p is the time-scaling factor.

The variable $l_{kw}(t)$ in (6.20) is defined as the state. It can be shown that the state satisfies the following difference equations (Zervos & Dumont, 1988)

$$\boldsymbol{L}_w(t+1)=\boldsymbol{K}_w\boldsymbol{L}_w(t)+\boldsymbol{H}_w a_w(t),\ w=1,\ldots,m.$$

where $\boldsymbol{L}_w(t)=[l_{1w},\cdots,l_{qw}]^T\in\mathbb{R}^q$, the matrices $\boldsymbol{K}_w\in\mathbb{R}^{q\times q}$ and $\boldsymbol{H}_w\in\mathbb{R}^q$ are defined in the following. If Δt is the sampling period and

$$\eta_1=e^{-p\Delta t},\eta_2=\Delta t+\frac{2}{p}(e^{-p\Delta t}-1),\eta_3=-\Delta te^{-p\Delta t}-\frac{2}{p}(e^{-p\Delta t}-1),\eta_4=\sqrt{2p}\frac{(1-e^{-p\Delta t})}{p},$$

then

$$\boldsymbol{K}_w=\begin{bmatrix}\eta_1 & 0 & \cdots & 0\\ \frac{-\eta_1\eta_2-\eta_3}{\Delta t} & \eta_1 & \cdots & 0\\ \vdots & \vdots & \vdots & \vdots\\ \frac{(-1)^{q-1}\eta_2^{q-2}(\eta_1\eta_2+\eta_3)}{\Delta t^{q-1}} & \cdots & \frac{-(\eta_1\eta_2+\eta_3)}{\Delta t} & \eta_1\end{bmatrix},$$

and

$$\boldsymbol{H}_w^T=\left[\eta_4,\ \ (-\eta_2/\Delta t)\eta_4,\ \ \cdots,\ \ (-\eta_2/\Delta t)^{q-1}\eta_4\right].$$

Finally, the state equation can be written as

$$\boldsymbol{L}(t+1)=\boldsymbol{K}\boldsymbol{L}(t)+\boldsymbol{H}\boldsymbol{a}(t), \tag{6.29}$$

where $\boldsymbol{L}(t)=[\boldsymbol{L}_1^T,\cdots,\boldsymbol{L}_m^T]^T\in\mathbb{R}^{mq}$, $\boldsymbol{K}=diag(\boldsymbol{K}_w)\in\mathbb{R}^{mq\times mq}$ and $\boldsymbol{H}=diag(\boldsymbol{H}_w)\in\mathbb{R}^{mq\times m}$.

The output equation can be derived as follows. Using (6.18), the model output is given by

$$y_n(x,t)=\sum_{i=1}^{n}\sum_{j_1=1}^{m}\sum_{k_1=1}^{q}\varphi_i(x)\theta_{ij_1k_1}^{(1)}\sum_{w_1=1}^{m}\psi_{j_1w_1}l_{k_1w_1}(t)+\sum_{i=1}^{n}\sum_{j_1=1}^{m}\sum_{j_2=1}^{m}\sum_{k_1=1}^{q}\sum_{k_2=1}^{q}\varphi_i(x)\theta_{ij_1j_2k_1k_2}^{(2)}$$
$$\times\sum_{w_1=1}^{m}\sum_{w_2=1}^{m}\psi_{j_1w_1}\psi_{j_2w_2}l_{k_1w_1}(t)l_{k_2w_2}(t)+\cdots.$$

Define $c_{k_1w_1}(x)=\sum_{i=1}^{n}\sum_{j_1=1}^{m}\varphi_i(x)\theta^{(1)}_{ij_1k_1}\psi_{j_1w_1}$ and $d_{k_1k_2w_1w_2}(x)=\sum_{i=1}^{n}\sum_{j_1=1}^{m}\sum_{j_2=1}^{m}\varphi_i(x)\theta^{(2)}_{ij_1j_2k_1k_2}\psi_{j_1w_1}\psi_{j_2w_2}$,

we have

$$y_n(x,t)=\sum_{w_1=1}^{m}\sum_{k_1=1}^{q}c_{k_1w_1}(x)l_{k_1w_1}(t)+\sum_{w_1=1}^{m}\sum_{w_2=1}^{m}\sum_{k_1=1}^{q}\sum_{k_2=1}^{q}d_{k_1k_2w_1w_2}(x)l_{k_1w_1}(t)l_{k_2w_2}(t)+\cdots.$$

Define

$$\boldsymbol{C}_{w_1}(x)=\begin{bmatrix}c_{1w_1}\\ \vdots\\ c_{qw_1}\end{bmatrix},\boldsymbol{D}_{w_1w_2}(x)=\begin{bmatrix}d_{11w_1w_2} & \cdots & d_{1qw_1w_2}\\ \vdots & \ddots & \vdots\\ d_{q1w_1w_2} & \cdots & d_{qqw_1w_2}\end{bmatrix},$$

then we have

$$y_n(x,t)=\sum_{w_1=1}^{m}[\boldsymbol{C}_{w_1}(x)^T\boldsymbol{L}_{w_1}(t)+\sum_{w_2=1}^{m}\boldsymbol{L}_{w_1}(t)^T\boldsymbol{D}_{w_1w_2}(x)\boldsymbol{L}_{w_2}(t)]+\cdots. \tag{6.30}$$

Finally, the output equation (6.30) can be further written in a simpler matrix form

$$y_n(x,t)=\boldsymbol{C}(x)^T\boldsymbol{L}(t)+\boldsymbol{L}(t)^T\boldsymbol{D}(x)\boldsymbol{L}(t)+\cdots, \tag{6.31}$$

where the spatial matrices $\boldsymbol{C}(x)$ and $\boldsymbol{D}(x)$ are given by

$$\boldsymbol{C}(x)=\begin{bmatrix}\boldsymbol{C}_1\\ \vdots\\ \boldsymbol{C}_m\end{bmatrix},\boldsymbol{D}(x)=\begin{bmatrix}\boldsymbol{D}_{11} & \cdots & \boldsymbol{D}_{1m}\\ \vdots & \ddots & \vdots\\ \boldsymbol{D}_{m1} & \cdots & \boldsymbol{D}_{mm}\end{bmatrix}.$$

It can be seen from (6.29) and (6.31) that the spatio-temporal Volterra model can be transformed into a spatio-temporal Wiener model in a state space form. Based on the state space model or original spatio-temporal Volterra model, traditional lumped controller design approaches such as model predictive control (Wang, 2004) and adaptive control (Zervos & Dumont, 1988) can be applied to the DPS.

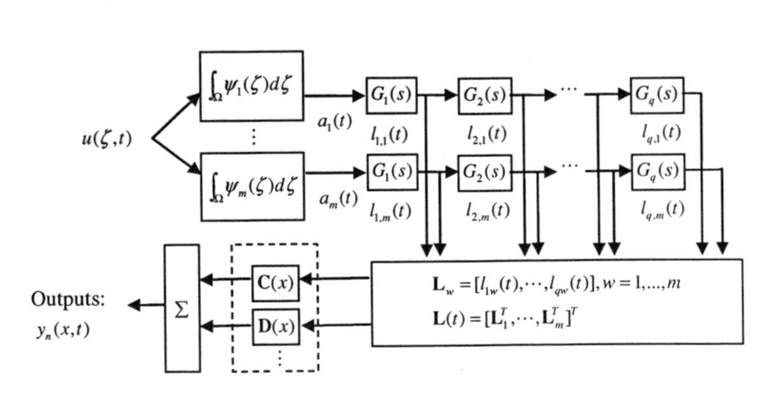

Fig. 6.2 Laguerre network for state space realization of Volterra model

6.5 Convergence Analysis

For simplicity, let $y_n(x,t)=V(x,t,\boldsymbol{\Theta},\{u(\zeta,\tau)\})$ denote a finite-order spatio-temporal Volterra model with R, n, $q<\infty$

$$\begin{aligned} y_n(x,t)=&\sum_{r=1}^{R}\int_{\Omega}\cdots\int_{\Omega}\sum_{\tau_1=0}^{t}\cdots\sum_{\tau_r=0}^{t}\sum_{i=1}^{n}\sum_{j_1=1}^{m}\cdots\sum_{j_r=1}^{m}\sum_{k_1=1}^{q}\cdots\sum_{k_r=1}^{q}\theta_{ij_1\ldots j_rk_1\ldots k_r}^{(r)}\\ &\times\varphi_i(x)\prod_{v=1}^{r}\psi_{j_v}(\zeta_v)\phi_{k_v}(\tau_v)u(\zeta_v,\tau_v)d\zeta_v, \end{aligned} \tag{6.32}$$

where the parameter matrix $\boldsymbol{\Theta}$ is defined by (6.25).

Because the functions $\{\psi_i(x)\}_{i=1}^{m}$, $\{\varphi_i(x)\}_{i=1}^{n}$ and $\{\phi_i(t)\}_{i=1}^{q}$ are basis functions, the spatio-temporal Volterra model structure $y_n(x,t)=V(x,t,\boldsymbol{\Theta},\{u(\zeta,\tau)\})$ is identifiable, i.e.,

$$V(x,t,\boldsymbol{\Theta}_1,\{u(\zeta,\tau)\})=V(x,t,\boldsymbol{\Theta}_2,\{u(\zeta,\tau)\})\Rightarrow\boldsymbol{\Theta}_1=\boldsymbol{\Theta}_2\,.$$

The convergence of the parameters and the model will be discussed below.

Based on the approximation result of Volterra series (Boyd & Chua, 1985) for traditional nonlinear processes, a nonlinear time-invariant DPS with the fading memory feature can also be approximated by a finite-order spatio-temporal Volterra model.

For a nonlinear time-invariant DPS $y(x,t)=N(\{u(\zeta,\tau)\})+d(x,t)$ with the fading memory feature on the input set U, and a finite-order spatio-temporal Volterra model $y_n(x,t)=V(x,t,\boldsymbol{\Theta},\{u(\zeta,\tau)\})$ with R, n, $q<\infty$, $\forall\delta>0$, we always assume that there exists a nonempty parameter set

$$D_{\boldsymbol{\Theta}}=\{\boldsymbol{\Theta}\,\|\,y(x,t)-V(x,t,\boldsymbol{\Theta},\{u(\zeta,\tau)\})|\le\delta,\forall u\in U\}\,.$$

Then there is an optimal model

$$y_n^*(x,t)=V(x,t,\boldsymbol{\Theta}^*,\{u(\zeta,\tau)\}), \tag{6.33}$$

with an optimal parameter matrix $\boldsymbol{\Theta}^*$ such that

$$\boldsymbol{\Theta}^*=\arg\min_{\boldsymbol{\Theta}\in D_{\boldsymbol{\Theta}}}\{\overline{E}(y(x,t)-V(x,t,\boldsymbol{\Theta},\{u(\zeta,\tau)\}))^2\}, \tag{6.34}$$

where $\overline{E}f(x,t)^2=\lim_{n_t\to\infty}\frac{1}{n_t}\sum_{t=1}^{n_t}\frac{1}{A}\int_{\Omega}Ef(x,t)^2dx$, $A=\int_{\Omega}dx$ and E is an expectation operator.

Under the uniform spatial discretization, $\frac{1}{A}\sum_{i=1}^{\infty}b_i(t)^2=\frac{1}{A}\int_{\Omega}y(x,t)^2dx$ can be replaced by $\frac{1}{n_y}\sum_{j=1}^{n_y}y(x_j,t)^2$. Therefore, according to the details of the identification algorithm developed in Section 6.3, the minimization problem (6.27) is indeed equivalent to the following problem

$$\hat{\Theta} = \arg\min_{\Theta \in D_\Theta} \{ \frac{1}{n_t} \frac{1}{n_y} \sum_{t=1}^{n_t} \sum_{j=1}^{n_y} (y(x_j,t) - V(x_j,t,\Theta,\{u(\zeta,\tau)\}))^2 \} . \tag{6.35}$$

It should be mentioned that the algorithm (6.27) can be considered as a practical implementation of (6.35) in order to reduce the involved spatial complexity. However, the theoretical analysis should be performed in the spatio-temporal domain.

Assumption 6.1:

For $y(x,t) = N(\{u(\zeta,\tau)\}) + d(x,t)$, let $W(x,t)$ be the σ-algebra generated by ($d(x,t), \cdots, d(x,0)$). For each t, τ ($t \ge \tau$) and any x, $\zeta \in \Omega$, there exist random variables $y_\tau^0(x,t)$ ($y_t^0(x,t) = 0$), $u_\tau^0(\zeta,t)$ ($u_t^0(\zeta,t) = 0$) that belong to $W(x,t)$ but are independent of $W(x,\tau)$, such that

$$E\,|\,y(x,t) - y_\tau^0(x,t)\,|^4 < M\lambda^{t-\tau},$$

$$E\,|\,u(\zeta,t) - u_\tau^0(\zeta,t)\,|^4 < M\lambda^{t-\tau},$$

for some $M < \infty$, $\lambda < 1$.

Assumption 6.2:

Assume that the model $y_n(x,t) = V(x,t,\Theta,\{u(\zeta,\tau)\})$ is differentiable with respect to Θ for all $\Theta \in D_\Theta$. Let D_Θ be compact. Assume that

$$|\,V(x,t,\Theta,\{u_1(\zeta,\tau)\}) - V(x,t,\Theta,\{u_2(\zeta,\tau)\})\,| \le M \sum_{\tau=0}^{t} \lambda^{t-\tau} \sup_{\zeta\in\Omega} |\,u_1(\zeta,\tau) - u_2(\zeta,\tau)\,|,$$

$$\left|\frac{\partial V(x,t,\Theta,\{u_1(\zeta,\tau)\})}{\partial \Theta} - \frac{\partial V(x,t,\Theta,\{u_2(\zeta,\tau)\})}{\partial \Theta}\right| \le M \sum_{\tau=0}^{t} \lambda^{t-\tau} \sup_{\zeta\in\Omega} |\,u_1(\zeta,\tau) - u_2(\zeta,\tau)\,|,$$

and $|\,V(x,t,\Theta,\{0(\zeta,\tau)\})\,| \le M$, where Θ belongs to an open neighborhood of D_Θ, $M < \infty$ and $\lambda < 1$.

Assumption 6.3:

Define $\varepsilon(x,t,\Theta) = y(x,t) - V(x,t,\Theta,\{u(\zeta,\tau)\})$ and there exists

$$\left|\frac{\partial \varepsilon(x,t,\Theta)^2}{\partial \Theta}\right| \le M\varepsilon(x,t,\Theta)^2,\ \Theta \in D_\Theta,\ \forall x \in \Omega,\ \forall t .$$

Theorem 6.1:

For a nonlinear time-invariant DPS $y(x,t) = N(\{u(\zeta,\tau)\}) + d(x,t)$ with the fading memory feature, let $\hat{y}_n(x,t) = V(x,t,\hat{\Theta},\{u(\zeta,\tau)\})$ be the spatio-temporal Volterra model where $\hat{\Theta}$ is estimated using (6.27). If Assumption 6.1, Assumption 6.2 and

Assumption 6.3 are satisfied, then $\hat{\Theta} \to \Theta^*$ and $\hat{y}_n(x,t) \to y_n^*(x,t)$ w. p. 1 as $n_t \to \infty$ and $n_y \to \infty$, where Θ^* and $y_n^*(x,t)$ are given by (6.34) and (6.33).

■

Proof:
In order to obtain the convergence with probability 1, the Lemma 5.2, which is the direct extension of the previous work (Cramér & Leadbetter, 1967; Ljung, 1978), is needed in the proof of Theorem 6.1.

We now turn to the proof of Theorem 6.1. Since $V(x,t,\Theta,\{u(\zeta,\tau)\})$ is continuous with respect to Θ, the convergence of parameters

$$\hat{\Theta} \to \Theta^*, \text{ w. p. 1 as } n_t \to \infty, n_y \to \infty, \tag{6.36}$$

naturally leads to the convergence of the model to its optimum

$$\hat{y}_n(x,t) \to y_n^*(x,t), \text{ w. p. 1 as } n_t \to \infty, n_y \to \infty.$$

Define

$$Q_{n_y n_t}(\Theta) = \frac{1}{n_y}\sum_{j=1}^{n_y}\{\frac{1}{n_t}\sum_{t=1}^{n_t}\varepsilon(x_j,t,\Theta)^2\}.$$

As defined in (6.34), Θ^* minimizes

$$\bar{E}(y(x,t) - V(x,t,\Theta,\{u(\zeta,\tau)\}))^2 = \lim_{n_y\to\infty}\{\lim_{n_t\to\infty} EQ_{n_y n_t}(\Theta)\},$$

and the estimate $\hat{\Theta}$ minimizes $Q_{n_y n_t}$ as defined in (6.35).

In order to prove (6.36), we should prove the following convergence

$$\sup_{\Theta\in D_\Theta} | Q_{n_y n_t}(\Theta) - EQ_{n_y n_t}(\Theta) | \to 0, \text{ w. p. 1 as } n_t \to \infty, n_y \to \infty. \tag{6.37}$$

One feasible solution is to achieve the following convergence at any fixed spatial variable x before working at the spatio-temporal space.

$$\sup_{\Theta\in D_\Theta} \frac{1}{n_t}\sum_{t=1}^{n_t} | \varepsilon(x,t,\Theta)^2 - E\varepsilon(x,t,\Theta)^2 | \to 0, \text{ w. p. 1 as } n_t \to \infty. \tag{6.38}$$

To achieve the convergence of (6.38), we have to obtain the convergence first at the pre-defined small open sphere, and then extend it to the global domain D_Θ using Heine-Borel's theorem.

Convergence of modeling error ε to its optimum over B:

Define the supremum between the model error and its optimum as a random variable

$$\eta(x,t)=\eta(x,t,\boldsymbol{\Theta}^0,\rho)=\sup_{\boldsymbol{\Theta}\in B}[\varepsilon(x,t,\boldsymbol{\Theta})^2-E\varepsilon(x,t,\boldsymbol{\Theta})^2].$$

Let D be the open neighborhood of $D_{\boldsymbol{\Theta}}$ and choose $\boldsymbol{\Theta}^0\in D_{\boldsymbol{\Theta}}$. We can define a small open sphere centered at $\boldsymbol{\Theta}^0$ as

$$B(\boldsymbol{\Theta}^0,\rho)=\{\boldsymbol{\Theta}\,|\,|\boldsymbol{\Theta}-\boldsymbol{\Theta}^0|<\rho\}.$$

Let $B=B(\boldsymbol{\Theta}^0,\rho)\cap D$, then

$$\sup_{\boldsymbol{\Theta}\in B}\frac{1}{n_t}\sum_{t=1}^{n_t}[\varepsilon(x,t,\boldsymbol{\Theta})^2-E\varepsilon(x,t,\boldsymbol{\Theta})^2]\le\frac{1}{n_t}\sum_{t=1}^{n_t}\eta(x,t).\tag{6.39}$$

Define $\xi(x,t)=\eta(x,t)-E\eta(x,t)$. *If we can prove*

- $\xi(x,t)$ *satisfies Lemma 5.2 and*
- *the mean of* $\eta(x,t)$ *is infinitesimal,*

then $\eta(x,t)$ *is also infinitesimal.*

Firstly, we consider

$$|E(\xi(x,t)\xi(x,\tau))|=Cov[\eta(x,t),\eta(x,\tau)].$$

Define $\eta_\tau^0(x,t)=\sup_{\boldsymbol{\Theta}\in B}[\varepsilon_\tau^0(x,t,\boldsymbol{\Theta})^2-E\varepsilon(x,t,\boldsymbol{\Theta})^2]$, with

$$\varepsilon_\tau^0(x,t,\boldsymbol{\Theta})=y_\tau^0(x,t)-V(x,t,\boldsymbol{\Theta},\{u_\tau^0(\zeta,j)\}),\ t>\tau,$$

where $\{u_\tau^0(\zeta,j)\}$ denotes the input set $(u_\tau^0(\zeta,t),...,u_\tau^0(\zeta,\tau+1),0,...,0)$ for all $\zeta\in\Omega$, $y_\tau^0(x,t)$ and $u_\tau^0(\zeta,j)$ are the variables introduced in Assumption 6.1. For convenience, let $u_\tau^0(\zeta,j)=0$ and $y_\tau^0(x,j)=0$ for $j<\tau$. Obviously $\eta_\tau^0(x,t)$ is independent of $\eta(x,\tau)$ from Assumption 6.1.

Hence

$$Cov[\eta(x,t),\eta(x,\tau)]=Cov[\eta(x,t)-\eta_\tau^0(x,t),\eta(x,\tau)].$$

Then using Schwarz's inequality, we have

$$|E(\xi(x,t)\xi(x,\tau))|\le[E\eta(x,\tau)^2E(\eta(x,t)-\eta_\tau^0(x,t))^2]^{1/2}.\tag{6.40}$$

Since

$$\begin{aligned}|\eta(x,t)-\eta_\tau^0(x,t)|&\le\sup_{\boldsymbol{\Theta}\in B}|\varepsilon(x,t,\boldsymbol{\Theta})^2-\varepsilon_\tau^0(x,t,\boldsymbol{\Theta})^2|\\&\le\sup_{\boldsymbol{\Theta}\in B}\{|\varepsilon(x,t,\boldsymbol{\Theta})|+|\varepsilon_\tau^0(x,t,\boldsymbol{\Theta})|\}\times\sup_{\boldsymbol{\Theta}\in B}|\varepsilon(x,t,\boldsymbol{\Theta})-\varepsilon_\tau^0(x,t,\boldsymbol{\Theta})|,\end{aligned}$$

using Assumption 6.2, we can further have

$$|\eta(x,t)-\eta_\tau^0(x,t)|\le M[\sum_{j=0}^{t}\lambda^{t-j}\{|y(x,j)|+|y_\tau^0(x,j)|+\sup_{\zeta\in\Omega}|u(\zeta,j)|+\sup_{\zeta\in\Omega}|u_\tau^0(\zeta,j)|\}]\times$$
$$[\sum_{j=0}^{t}\lambda^{t-j}\{|y(x,j)-y_\tau^0(x,j)|+\sup_{\zeta\in\Omega}|u(\zeta,j)-u_\tau^0(\zeta,j)|\}].$$

Using Assumption 6.1 and Schwarz's inequality, we can finally derive

$$E|\eta(x,t)-\eta_\tau^0(x,t)|^2\le M\lambda^{t-\tau}. \tag{6.41}$$

Following the similar derivation above and using Assumption 6.2 and Assumption 6.1, we can also derive

$$E\eta(x,\tau)^2\le M. \tag{6.42}$$

Placing (6.41) and (6.42) into (6.40), we can easily derive that $\xi(x,t)$ satisfies Lemma 5.2, that is

$$\frac{1}{n_t}\sum_{t=1}^{n_t}\xi(x,t)=\frac{1}{n_t}\sum_{t=1}^{n_t}(\eta(x,t)-E\eta(x,t))\to 0,\ \text{w. p. 1 as } n_t\to\infty. \tag{6.43}$$

Secondly, the mean value of η can be expressed as,

$$E\eta(x,t)=E\sup_{\boldsymbol{\Theta}\in B}[\varepsilon(x,t,\boldsymbol{\Theta})^2-E\varepsilon(x,t,\boldsymbol{\Theta})^2].$$

Since the right-hand side is continuous with respect to $\boldsymbol{\Theta}$, $E\eta(x,t)$ should be small if B is small. Furthermore, by Assumption 6.3,

$$|\frac{\partial\varepsilon(x,t,\boldsymbol{\Theta})^2}{\partial\boldsymbol{\Theta}}|\le M|\varepsilon(x,t,\boldsymbol{\Theta})|^2\le M[\sum_{j=0}^{t}\lambda^{t-j}\{|y(x,j)|+\sup_{\zeta\in\Omega}|u(\zeta,j)|\}]^2,$$

where we again have used the uniform bounds in Assumption 6.2. Consequently, by Assumption 6.1,

$$E\sup_{\boldsymbol{\Theta}\in B}|\frac{\partial\varepsilon(x,t,\boldsymbol{\Theta})^2}{\partial\boldsymbol{\Theta}}|^2\le M.$$

Now

$$\begin{aligned}E\eta(x,t)&=E\sup_{\boldsymbol{\Theta}\in B}[\varepsilon(x,t,\boldsymbol{\Theta})^2-E\varepsilon(x,t,\boldsymbol{\Theta})^2]\\&\le E\sup_{\boldsymbol{\Theta}\in B}[\varepsilon(x,t,\boldsymbol{\Theta})^2-\varepsilon(x,t,\boldsymbol{\Theta}^0)^2]+\sup_{\boldsymbol{\Theta}\in B}E[\varepsilon(x,t,\boldsymbol{\Theta}^0)^2-\varepsilon(x,t,\boldsymbol{\Theta})^2]\\&\le[E\sup_{\boldsymbol{\Theta}\in B}|\frac{\partial\varepsilon(x,t,\boldsymbol{\Theta})^2}{\partial\boldsymbol{\Theta}}|+\sup_{\boldsymbol{\Theta}\in B}E|\frac{\partial\varepsilon(x,t,\boldsymbol{\Theta})^2}{\partial\boldsymbol{\Theta}}|]\times\sup_{\boldsymbol{\Theta}\in B}|\boldsymbol{\Theta}-\boldsymbol{\Theta}^0|\\&\le M^0\rho.\end{aligned} \tag{6.44}$$

Finally from (6.44), (6.39) becomes

$$\sup_{\boldsymbol{\Theta}\in B}\frac{1}{n_t}\sum_{t=1}^{n_t}[\varepsilon(x,t,\boldsymbol{\Theta})^2-E\varepsilon(x,t,\boldsymbol{\Theta})^2]\le\frac{1}{n_t}\sum_{t=1}^{n_t}(\eta(x,t)-E\eta(x,t))+M^0\rho. \tag{6.45}$$

It is clearly seen from (6.43) that the first term of the right-hand side is arbitrarily small for a sufficiently large n_t. Since ρ can be arbitrarily small, therefore

$$\sup_{\Theta\in B}\frac{1}{n_t}\sum_{t=1}^{n_t}|\varepsilon(x,t,\Theta)^2 - E\varepsilon(x,t,\Theta)^2| \to 0,\text{ w. p. 1 as } n_t \to \infty. \tag{6.46}$$

Convergence extension to global D_Θ***:***

Since D_Θ is compact, by applying Heine-Borel's theorem, the following result is easily derived from (6.46),

$$\sup_{\Theta\in D_\Theta}\frac{1}{n_t}\sum_{t=1}^{n_t}|\varepsilon(x,t,\Theta)^2 - E\varepsilon(x,t,\Theta)^2| \to 0,\text{ w. p. 1 as } n_t \to \infty. \tag{6.47}$$

Extension to spatio-temporal domain:

Obviously, we have

$$\sup_{\Theta\in D_\Theta}|Q_{n_y n_t}(\Theta) - EQ_{n_y n_t}(\Theta)| \le \frac{1}{n_y}\sum_{j=1}^{n_y}\sup_{\Theta\in D_\Theta}\frac{1}{n_t}\sum_{t=1}^{n_t}|\varepsilon(x_j,t,\Theta)^2 - E\varepsilon(x_j,t,\Theta)^2|,$$

Thus, the following conclusion is derived,

$$\sup_{\Theta\in D_\Theta}|Q_{n_y n_t}(\Theta) - EQ_{n_y n_t}(\Theta)| \to 0,\text{ w. p. 1 as } n_t \to \infty,\ n_y \to \infty. \tag{6.48}$$

■

6.6 Simulation and Experiment

In order to evaluate the presented modeling method, the simulation on the catalytic rod is studied first. Then the experiment and modeling for snap curing oven are presented. For an easy comparison, some performance indexes are established for the DPS as follows

- Spatial normalized absolute error, $SNAE(t) = \int |e(x,t)|dx / \int dx$,
- Temporal normalized absolute error, $TNAE(x) = \sum |e(x,t)| / \sum \Delta t$,
- Root of mean squared error, $RMSE = (\int \sum e(x,t)^2 dx / \int dx \sum \Delta t)^{1/2}$.

6.6.1 Catalytic Rod

Consider the catalytic rod given in Sections 1.1.2 and 3.6.1. In the simulation, assume the process noise $d(x,t)$ in (3.31) is zero. Twenty-two sensors uniformly distributed in the space are used to measure the temperature distribution. The sampling interval is $\Delta t = 0.01$ and the simulation time is 5.

The measured process output $y(x,t)$ is shown in Figure 6.3, where the first 250 data are used for model training, the next 150 data for validation and the remaining 100 data for model testing. As show in Figure 6.4, the first four Karhunen-Loève basis functions are used for the spatio-temporal Volterra modeling. The temporal bases $\phi_i(t)$ are chosen as Laguerre functions with the time-scaling factor $p = 4.5$ and the truncation length $q = 4$ using the cross-validation method.

The predicted output $\hat{y}_n(x,t)$ and the prediction error $e(x,t) = y - \hat{y}_n$ of the *2*nd*-order spatio-temporal Volterra model* are presented in Figure 6.5 and Figure 6.6 respectively. It is obvious that the *2*nd*-order spatio-temporal Volterra model* can satisfactorily model the process. For many applications, the 2nd-order model is enough for a good approximation, and too high-order models may cause the over-complexity problem. As shown in Figure 6.7 and Figure 6.8, the *2*nd*-order spatio-temporal Volterra model* performs much better than the *1*st*-order spatio-temporal Volterra model* (i.e., *Green's function model*) because the 1st-order spatio-temporal Volterra model is basically a linear system.

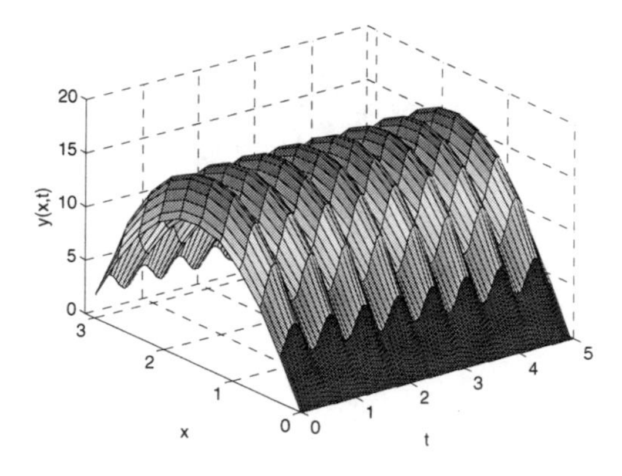

Fig. 6.3 Catalytic rod: Measured output for Volterra modeling

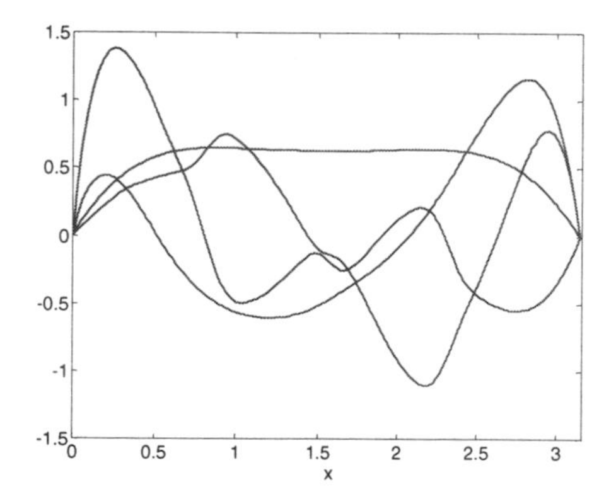

Fig. 6.4 Catalytic rod: KL basis functions for Volterra modeling

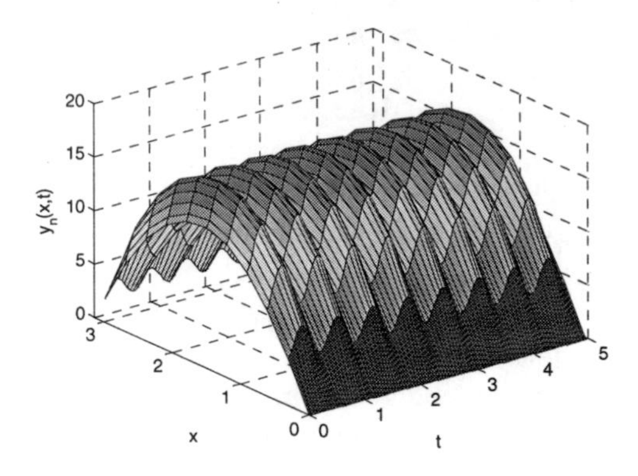

Fig. 6.5 Catalytic rod: Predicted output of 2nd-order Volterra model

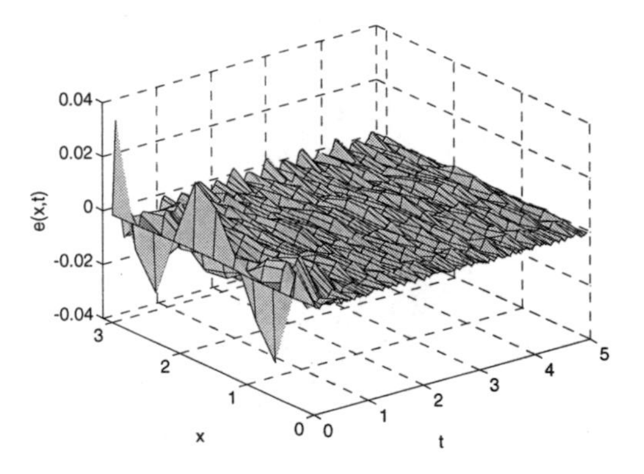

Fig. 6.6 Catalytic rod: Prediction error of 2nd-order Volterra model

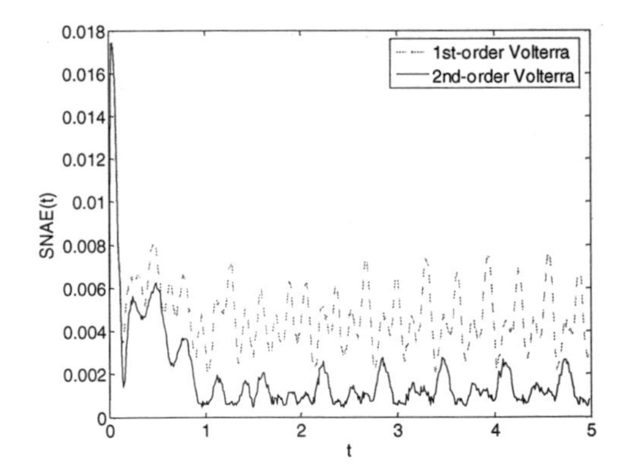

Fig. 6.7 Catalytic rod: $SNAE$(t) of 1st and 2nd-order Volterra models

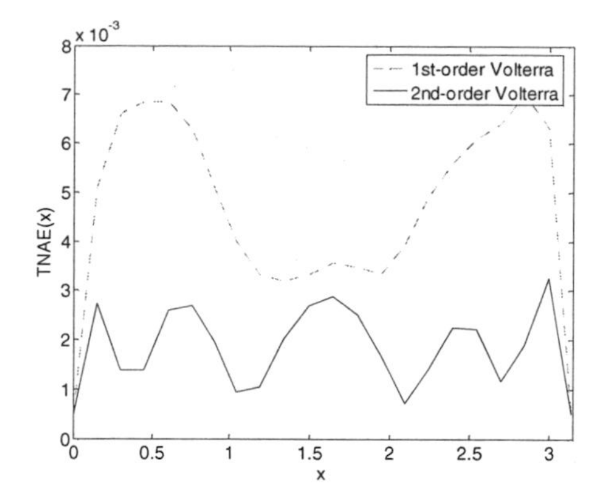

Fig. 6.8 Catalytic rod: *TNAE*(x) of 1st and 2nd-order Volterra models

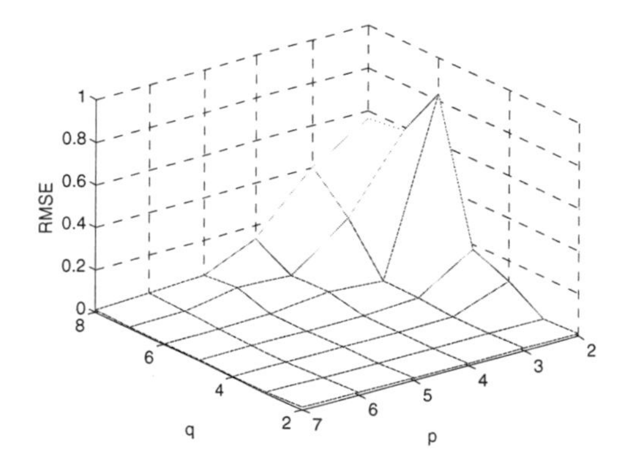

Fig. 6.9 Catalytic rod: *RMSE* of 2nd-order Volterra model

The modeling performance is also affected by the number of Laguerre temporal basis functions (q) and the time-scaling factor (p).Figure 6.9 displays *RMSE* of the 2nd-*order spatio-temporal Volterra model* with respect to these two parameters. It is shown that this modeling approach is robust to these parameters since there are a wide range of parameters which can be chosen to obtain a good performance.

6.6.2 Snap Curing Oven

Consider the snap curing oven (Figure 1.1 and Figure 3.11) provided in Sections 1.1.1 and 3.6.2. In the experiment, a total of 2100 measurements are collected with a sampling interval $\Delta t = 10$ seconds. One thousand and four hundred of measurements from sensors (s1-s5, s7-s10, and s12-s16) are used to estimate the model. The last 700 measurements from sensors (s1-s5, s7-s10, and s12-s16) are chosen to validate the model and determine the time-scaling factor and truncation length of

Laguerre series using the cross-validation method. All 2100 measurements from the rest sensors (s6, s11) are used for model testing.

In the spatio-temporal Volterra modeling, five two-dimensional Karhunen-Loève basis functions are used as spatial bases and the first two of them are shown in Figure 6.10 and Figure 6.11. The temporal bases $\phi_i(t)$ are chosen as Laguerre series with the time-scaling factor $p = 0.001$ and the truncation length $q = 3$.

The 2nd*-order spatio-temporal Volterra model* is used to model the thermal process. After the training using the first 1400 data from the sensors (s1-s5, s7-s10, and s12-s16), a process model can be obtained with the significant performance such as the sensor s1 in Figure 6.12. The model also performs very well for the untrained locations such as the sensor s6 in Figure 6.13. The predicted temperature distribution of the oven at t=10000s is provided in Figure 6.14. The performance

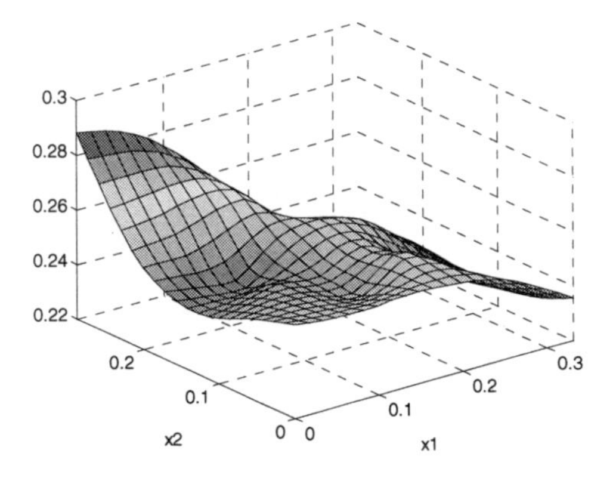

Fig. 6.10 Snap curing oven: KL basis functions (i=1) for Volterra modeling

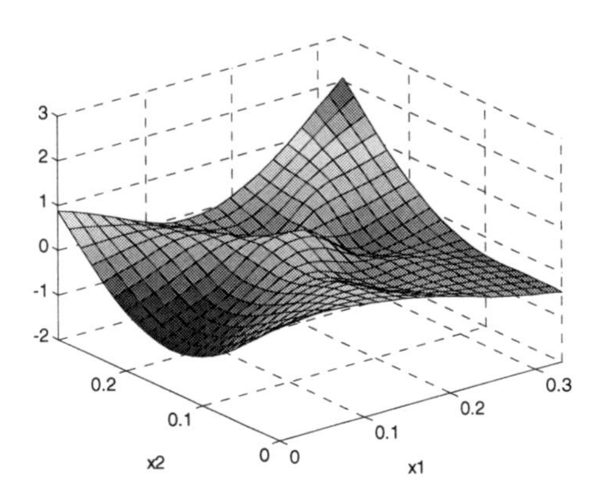

Fig. 6.11 Snap curing oven: KL basis functions (i=2) for Volterra modeling

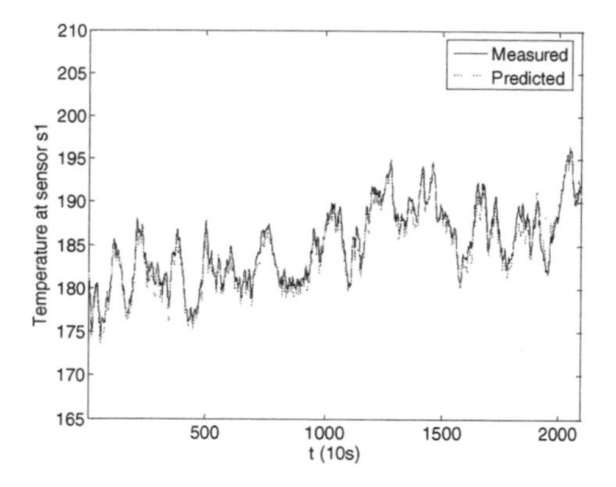

Fig. 6.12 Snap curing oven: Performance of 2nd-order Volterra model at sensor s1

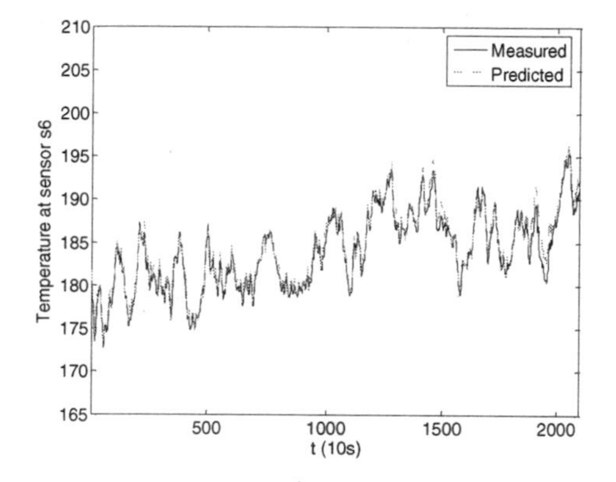

Fig. 6.13 Snap curing oven: Performance of 2nd-order Volterra model at sensor s6

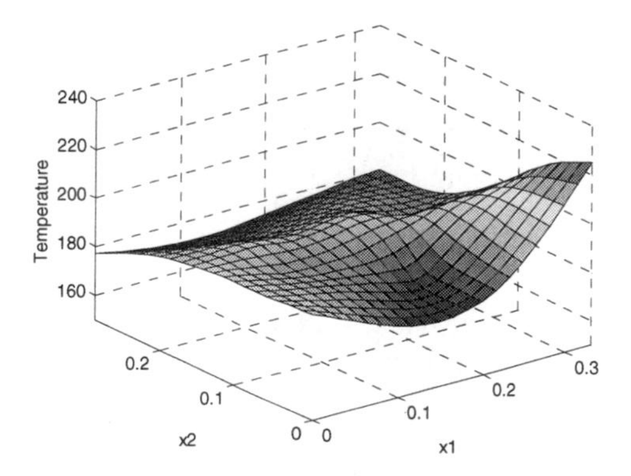

Fig. 6.14 Snap curing oven: Predicted temperature distribution of 2nd-order Volterra model at t=10000s

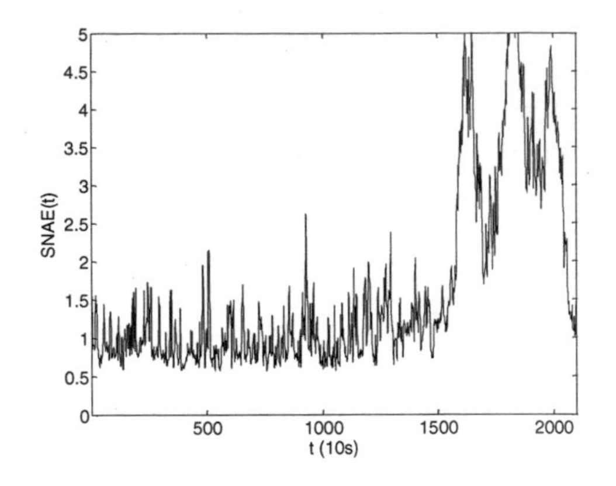

Fig. 6.15 Snap curing oven: *SNAE*(t) of 1st-order Volterra model

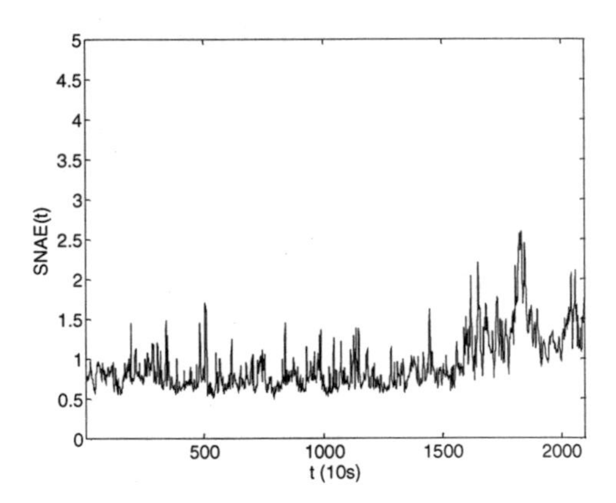

Fig. 6.16 Snap curing oven: *SNAE*(t) of 2nd-order Volterra model

Table 6.1 Snap curing oven: *TNAE*(x) of 1st and 2nd-order Volterra models

	s1	s2	s3	s4	s5	s6	s7	s8
1st-order model	1.87	1.44	1.75	1.34	1.75	1.55	1.5	1.59
2nd-order model	0.79	0.71	1.2	0.71	1.59	0.65	0.78	1.24
	s9	s10	s11	s12	s13	s14	s15	s16
1st-order model	1.13	1.81	2.1	1.43	1.85	2.22	1.13	1.52
2nd-order model	0.86	0.86	1.33	1.08	0.71	1	0.7	0.55

comparisons over the whole data set in Figure 6.15 and Figure 6.16 and Table 6.1 further show that the *2nd-order spatio-temporal Volterra model* has a much better performance than the *1st-order spatio-temporal Volterra model*. The effectiveness of the presented modeling method is clearly demonstrated in this real application.

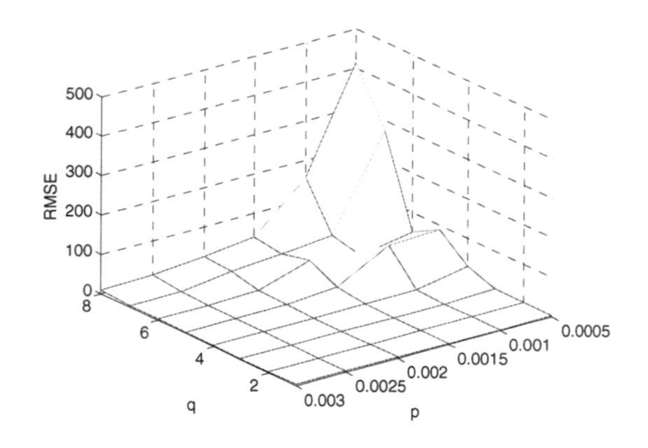

Fig. 6.17 Snap curing oven: *RMSE* of 2nd-order Volterra model

As shown in Figure 6.17, a good modeling performance can be maintained in a wide range of parameter space about the number of Laguerre temporal basis functions (q) and the time-scaling factor (p). Note that though the modeling gives some certain robustness to the selection of p and q, their selection is still very important in modeling of real systems. If they are not properly selected, the modeling algorithm may become divergent due to the process noise and disturbance.

6.7 Summary

A Volterra kernel based spatio-temporal modeling approach is presented for unknown nonlinear distributed parameter systems. The spatio-temporal Volterra model is constructed, where the kernels are functions of time and space variables. In order to estimate the kernels, they are expanded onto Karhunen-Loève spatial bases as well as Laguerre temporal bases with unknown coefficients. With the help of the Galerkin method, these unknown parameters can be estimated from the process data using the least-squares method in the temporal domain. This spatio-temporal Volterra model can achieve a better performance than the Green's function model. The presented modeling algorithm provides convergent and satisfactory estimates. The simulation and experiment are conducted to demonstrate the effectiveness of the presented modeling method and its potential application to a wide range of nonlinear DPS.

References

[1] Boyd, S., Chua, L.O.: Fading memory and the problem of approximating nonlinear operators with Volterra series. IEEE Transactions on Circuits and Systems 32(11), 1150–1161 (1985)

[2] Butkovskiy, A.G.: Green's functions and transfer functions handbook, 1st edn. Ellis Horwood, Chichester (1982)

[3] Christofides, P.D.: Nonlinear and robust control of PDE systems: Methods and applications to transport-reaction processes. Birkhäuser, Boston (2001b)
[4] Cramér, H., Leadbetter, M.R.: Stationary and related stochastic processes. Wiely, New York (1967)
[5] Datta, K.B., Mohan, B.M.: Orthogonal functions in systems and control. World Scientific, Singapore (1995)
[6] Doumanidis, C.C., Fourligkas, N.: Temperature distribution control in scanned thermal processing of thin circular parts. IEEE Transactions on Control Systems Technology 9(5), 708–717 (2001)
[7] Doyle III, F.J., Ogunnaike, B.A., Pearson, R.K.: Nonlinear model-based control using second-order Volterra Models. Automatica 31(5), 697–714 (1995)
[8] Gay, D.H., Ray, W.H.: Identification and control of distributed parameter systems by means of the singular value decomposition. Chemical Engineering Science 50(10), 1519–1539 (1995)
[9] Heuberger, P.S.C., Van den Hof, P.M.J., Bosgra, O.H.: A generalized orthonormal basis for linear dynamical systems. IEEE Transactions on Automatic Control 40(3), 451–465 (1995)
[10] Ljung, L.: Convergence analysis of parametric identification methods. IEEE Transactions on Automatic Control 23(5), 770–783 (1978)
[11] Ljung, L.: System identification: Theory for the user, 2nd edn. Prentice-Hall, Englewood Cliffs (1999)
[12] Maner, B.R., Doyle III, F.J., Ogunnaike, B.A., Pearson, R.K.: Nonlinear model predictive control of a simulated multivariable polymerization reactor using second-order Volterra models. Automatica 32(9), 1285–1301 (1996)
[13] Park, H.M., Cho, D.H.: Low dimensional modeling of flow reactors. International Journal of Heat and Mass Transfer 39(16), 3311–3323 (1996a)
[14] Park, H.M., Cho, D.H.: The use of the Karhunen-Loève decomposition for the modeling of distributed parameter systems. Chemical Engineering Science 51(1), 81–98 (1996b)
[15] Parker, R.S., Heemstra, D., Doyle III, F.J., Pearson, R.K., Ogunnaike, B.A.: The identification of nonlinear models for process control using tailored "plant-friendly" input sequences. Journal of Process Control 11(2), 237–250 (2001)
[16] Rugh, W.: Nonlinear system theory: The Volterral/Wiener approach. Johns Hopkins University Press, Baltimore (1981)
[17] Schetzen, M.: The Volterra and Wiener theories of nonlinear systems. Wiley, New York (1980)
[18] Wahlberg, B.: System identification using Laguerre models. IEEE Transactions on Automatic Control 36(5), 551–562 (1991)
[19] Wahlberg, B.: System identification using Kautz models. IEEE Transactions on Automatic Control 39(6), 1276–1282 (1994)
[20] Wahlberg, B., Mäkilä, P.M.: Approximation of stable linear dynamical systems using Laguerre and Kautz functions. Automatica 32(5), 693–708 (1996)
[21] Wang, L.: Discrete model predictive controller design using Laguerre functions. Journal of Process Control 14(2), 131–142 (2004)
[22] Zervos, C.C., Dumont, G.A.: Deterministic adaptive control based on Laguerre series representation. International Journal of Control 48(6), 2333–2359 (1988)

[23] Zheng, D., Hoo, K.A.: Low-order model identification for implementable control solutions of distributed parameter system. Computers and Chemical Engineering 26(7-8), 1049–1076 (2002)
[24] Zheng, D., Hoo, K.A.: System identification and model-based control for distributed parameter systems. Computers and Chemical Engineering 28(8), 1361–1375 (2004)
[25] Zheng, D., Hoo, K.A., Piovoso, M.J.: Low-order model identification of distributed parameter systems by a combination of singular value decomposition and the Karhunen-Loève expansion. Industrial & Engineering Chemistry Research 41(6), 1545–1556 (2002)

7 Nonlinear Dimension Reduction Based Neural Modeling for Nonlinear Complex DPS

Abstract. A nonlinear principal component analysis (NL-PCA) based neural modeling approach is presented for a lower-order or more accurate solution for nonlinear distributed parameter systems (DPS). One NL-PCA network is trained for the nonlinear dimension reduction and the nonlinear time/space reconstruction. The other neural model is to learn the system dynamics with a linear/nonlinear separated model structure. With the powerful capability of dimension reduction and the intelligent learning, this approach can model the nonlinear complex DPS with much more complexity. The simulation on the catalytic rod and the experiment on the snap curing oven will demonstrate the effectiveness of the presented method.

7.1 Introduction

Karhunen-Loève (KL) decomposition has been widely used for distributed parameter system (DPS) identification with the help of traditional system identification techniques (Sahan *et al.*, 1997; Zhou, Liu, Dai & Yuan, 1996; Aggelogiannaki & Sarimveis, 2007; Smaoui & Al-Enezi, 2004; Qi & Li, 2008a). Karhunen-Loève decomposition, also called principal component analysis (PCA) (Baker & Christofides, 2000; Armaou & Christofides, 2002; Park & Cho, 1996a, 1996b; Hoo & Zheng, 2001; Newman, 1996a, 1996b), is a popular approach to find the principal spatial structures from the data. Among all linear expansion, PCA basis functions can give a lower-dimensional model. However, the traditional PCA is a linear dimension reduction (i.e., linear projection and linear reconstruction) method, and it may not be very suitable to model the nonlinear dynamics efficiently (Malthouse, 1998). This is because PCA produces a linear approximation to the original nonlinear problem, which may not guarantee the assumption that minor components do not contain important information (Wilson, Irwin & Lightbody, 1999).

This has naturally motivated the development of nonlinear PCA (NL-PCA) for the nonlinear problem. NL-PCA is a nonlinear dimension reduction method, which can retain more information using fewer components. In the field of machine learning, NL-PCA has been used to deal with the nonlinear dimension reduction problem. Examples include principal curves/surfaces (Dong & McAvoy, 1996), multi-layer auto-associative neural networks (Kramer, 1991; Saegusa, Sakano & Hashimoto, 2004; Hsieh, 2001; Hinton & Salakhutdinov, 2006; Kirby & Miranda, 1994; Smaoui, 2004), the kernel function approach (Webb, 1996; Schölkopf, Smola & Muller, 1998), and the radial basis function (RBF) networks (Wilson, Irwin &

H.-X. Li and C. Qi: Spatio-Temporal Modeling of Nonlinear DPS, ISCA 50, pp. 149–165.
springerlink.com

Lightbody, 1999). However, they mainly focus on the reduction and analysis of the high-dimensional data or the known system.

In this chapter, a nonlinear dimension reduction based spatio-temporal modeling is developed for unknown nonlinear distributed parameter systems. The approach consists of two major stages. First, a NL-PCA network is trained for the nonlinear dimension reduction. With the help of finite spatial measurement, the well-trained NL-PCA network can transform and reduce the high-dimensional spatio-temporal data into the low-dimensional time series data and reconstruct them back to the spatio-temporal data. Then, a low-order neural network can be easily established for dynamic modeling with the help of traditional identification technique. The simulation and experiment demonstrates that this nonlinear dimension reduction based modeling can achieve a better performance than the linear dimension reduction based approach when modeling the nonlinear DPS.

This chapter is organized as follows. The novel NL-PCA based spatio-temporal modeling methodology is described in Section 7.2. The presented NL-PCA based spatio-temporal modeling is implemented using neural networks in Section 7.3, with simulation and experiment conducted in Section 7.4. Finally, a few conclusions are presented in Section 7.5.

7.2 Nonlinear PCA Based Spatio-Temporal Modeling Framework

7.2.1 Modeling Methodology

Consider a nonlinear distributed parameter system, $u(t)\in\mathbb{R}^m$ is the temporal input, and $y(x,t)\in\mathbb{R}$ is the spatio-temporal output, where $x\in\Omega$ is the spatial variable, Ω is the space domain, and t is the temporal variable. Here the system is controlled by the m actuators with temporal signal $u(t)$ and certain spatial distribution. The output is measured at the N spatial locations $x_1,\ldots,x_N$. The modeling objective is to identify a low-order spatio-temporal model from the temporal input $\{u(t)\}_{t=1}^{L}$ and the spatio-temporal output $\{y(x_i,t)\}_{i=1,t=1}^{N,L}$, where L is the time length.

One method, called "modeling-then-reduce", is to identify a high-order model from the data and then perform the model reduction. For the high-dimensional model, a large number of functions and parameters need to be estimated, and a complex model reduction has to be used to derive a low-order model.

In this study, a two-stage spatio-temporal modeling methodology is used and depicted in Figure 7.1. The first stage is for the dimension reduction, where the high-dimensional spatio-temporal output data $\{y(x_i,t)\}_{i=1,t=1}^{N,L}$ can be reduced to the low-dimensional time series data $\{y(t)\}_{t=1}^{L}$. The second stage is for the dynamic modeling, where a low-order temporal model is identified using the time series data. Using the prediction of the temporal model and the nonlinear time/space reconstruction, this modeling method can reproduce the spatio-temporal dynamics of the distributed system. This methodology can be regarded as "reduce-then-modeling".

It can greatly reduce the computational complexity involved in the low-order modeling.

As shown in Figure 7.1(a), the nonlinear dimension reduction approach should include two functions:

- Nonlinear projection for time/space separation - a nonlinear transformation from the high-dimensional time/space domain into the low-dimensional time series.
- Nonlinear time/space reconstruction - a nonlinear reverse transformation from the reduced-dimensional time series to the original time/space domain.

Obviously, the projection and the reconstruction should be invertible for a good reproduction of the original spatio-temporal dynamics. As shown in Figure 7.1(b), the first step is to learn the nonlinear projection and reconstruction using NL-PCA method, and the second is to model the system dynamics using traditional nonlinear identification techniques.

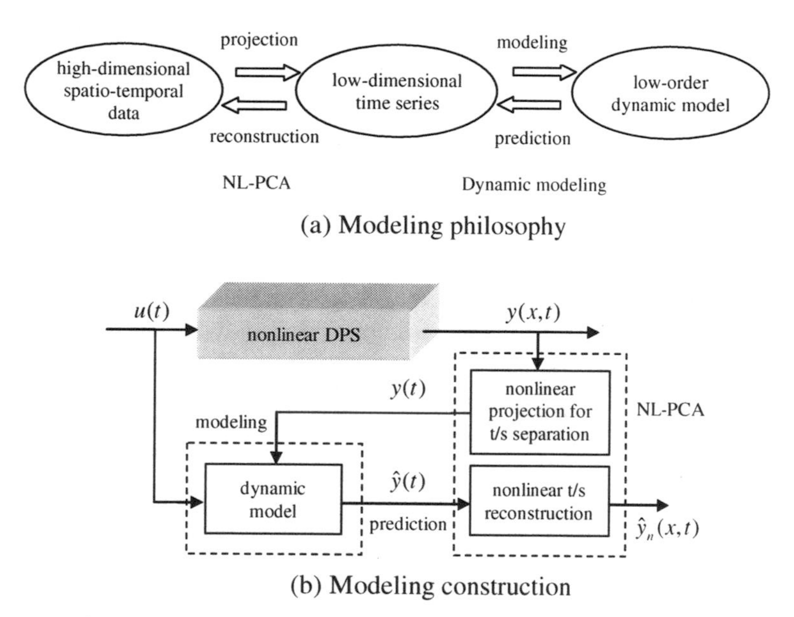

Fig. 7.1 NL-PCA based spatio-temporal modeling methodology

7.2.2 *Principal Component Analysis*

The main problem of using PCA for the dimension reduction is to compute the most characteristic spatial structure $\{\varphi_i(x)\}_{i=1}^{n}$ among the spatio-temporal output $\{y(x_i,t)\}_{i=1,t=1}^{N,L}$, where n is the number of spatial basis functions (i.e., the dimension of the time domain, $n<N$). For simplicity, assume the process output is uniformly sampled in time and space.

Taking $(\varphi_i, \varphi_i) = 1$ without loss of generality, the projection and reconstruction of PCA have the following linear form

$$y_i(t) = (\varphi_i(x), y(x,t)), i = 1,...,n \,, \tag{7.1}$$

$$y_n(x,t) = \sum_{i=1}^{n} \varphi_i(x) y_i(t) \,. \tag{7.2}$$

Then the reconstruction error can be defined as $e(x,t) = y(x,t) - y_n(x,t)$. The spatial principal axes $\{\varphi_i(x)\}_{i=1}^{n}$ can be found by minimizing the following objective function

$$\min_{\varphi_i(x)} \; <\| e(x,t) \|^2> . \tag{7.3}$$

Spatial correlation method

There are several implementations of PCA. The simple one is based on the spatial correlation function. The necessary condition of the solution of (7.3) can be obtained as below (Holmes, Lumley & Berkooz, 1996)

$$\int_{\Omega} R(x,\zeta)\varphi_i(\zeta) d\zeta = \lambda_i \varphi_i(x), \; (\varphi_i, \varphi_i) = 1, \; i = 1,...,n \,, \tag{7.4}$$

where $R(x,\zeta) =< y(x,t) y(\zeta,t) >$ is the spatial two-point correlation function. Since the matrix R is symmetric and positive semidefinite, the computed eigenfunctions are orthogonal.

Because the data are always discrete in space, one must solve numerically the integral equation (7.4). Discretizing the integral equation gives a $N \times N$ matrix eigenvalue problem. Thus, at most N eigenfunctions at N sampled spatial locations can be obtained. Then one can interpolate the eigenfunctions to locations where the data are not available.

Dimension determination

The maximum number of nonzero eigenvalues is $K \le N$. Let the eigenvalues $\lambda_1 > \lambda_2 > \cdots > \lambda_K$ and the corresponding eigenfunctions $\varphi_1(x)$, $\varphi_2(x)$, $\cdots$, $\varphi_K(x)$ in the order of the magnitude of the eigenvalues. The ratio of the sum of the n largest eigenvalues to the total sum

$$\eta = \sum_{i=1}^{n} \lambda_i / \sum_{i=1}^{K} \lambda_i \,, \tag{7.5}$$

gives the fraction of the variance retained in the n-dimensional space defined by the eigenfunctions associated with those eigenvalues (Webb, 1996). One method of estimating the intrinsic dimension is to take the value of n for which η is sufficiently large. A value of more than 0.9 is often taken. In general, only the first few basis functions expansion can represent the dominant dynamics of the spatio-temporal system. It can be shown (Holmes, Lumley & Berkooz, 1996) that for some arbitrary set of basis functions $\{\phi_i(x)\}_{i=1}^n$, the following result holds

$$\sum_{i=1}^{n} <(y(\cdot,t),\varphi_i)^2> = \sum_{i=1}^{n} \lambda_i^2 \geq \sum_{i=1}^{n} <(y(\cdot,t),\phi_i)^2> .$$

It means that the PCA expansion is optimal on average in the class of linear dimension reduction methods.

7.2.3 Nonlinear PCA for Projection and Reconstruction

Unlike the linear PCA, the NL-PCA allows the projection and the reconstruction to be arbitrary nonlinear function. Finding these two functions is the first step for the spatio-temporal modeling.

The nonlinear projection is in the form

$$y_i(t) = G_i(y_x(t)), i = 1,...,n , \tag{7.6}$$

where $y_x(t) = [y(x_1,t),..., y(x_N,t)]^T \in \mathbb{R}^N$ and n is the reduced dimension in the time domain ($n < N$). A compact vector form is described in the following

$$y(t) = G(y_x(t)) , \tag{7.7}$$

where $y(t) = [y_1(t),\cdots, y_n(t)]^T \in \mathbb{R}^n$ and $G(\cdot) = [G_1(\cdot),\cdots,G_n(\cdot)]^T : \mathbb{R}^N \to \mathbb{R}^n$. Here $G_i(\cdot)$: $\mathbb{R}^N \to \mathbb{R}$ is referred to as the ith nonlinear spatial principal axis.

The reconstruction is completed by a second nonlinear function $H(\cdot): \mathbb{R}^{n+1} \to \mathbb{R}$ as below

$$y_n(x,t) = H(x, y(t)) . \tag{7.8}$$

Here $y_n(x,t)$ denotes the nth-order approximation, and the function H with the spatial variable x is used to reproduce the output at sampled spatial locations as well as any other locations.

The reconstruction error is measured by $e(x,t) = y(x,t) - y_n(x,t)$. The functions G and H are selected to minimize the following objective function

$$\min_{G,H} <\| e(x,t) \|^2> . \tag{7.9}$$

where the ensemble average, norm and inner product are defined as $< f(x,t) >= \frac{1}{L}\sum_{t=1}^{L} f(x,t)$, $\| f(x) \|= (f(x), f(x))^{1/2}$ and $(f(x), g(x)) = \int_\Omega f(x)g(x)dx$.

7.2.4 Dynamic Modeling

The second step for the spatio-temporal modeling is to establish the dynamic relationship between $y(t)$ and $u(t)$. Due to the reduced dimensionality of $y(t)$, the spatio-temporal modeling becomes a simple low-order temporal modeling, for which traditional nonlinear system identification techniques can be easily applied.

The time series data $y(t)$ is often described by a deterministic nonlinear auto-regressive with exogenous input (NARX) model (Leontaritis & Billings, 1985)

$$y(t) = F(y(t-1),\cdots,y(t-d_y),u(t-1),\cdots,u(t-d_u)) + e(t)\,, \tag{7.10}$$

where d_u and d_y denote the maximum input and output lags respectively, and $e(t)$ denotes the model error. The unknown function F can be estimated from the input/output data $\{u(t), y(t)\}_{t=1}^{L}$ using such as polynomials, radial basis functions, wavelets and kernel functions (Sjöberg *et al.*, 1995), where the low-dimensional time series $y(t)$ can be obtained using the nonlinear projection function of NL-PCA described in (7.7).

In summary, the estimated spatio-temporal model is composed of a low-order temporal model

$$\hat{y}(t) = \hat{F}(\hat{y}(t-1),\cdots,\hat{y}(t-d_y),u(t-1),\cdots,u(t-d_u))\,, \tag{7.11}$$

and a spatio-temporal reconstruction equation

$$\hat{y}_n(x,t) = H(x,\hat{y}(t))\,, \tag{7.12}$$

where $\hat{y}(t)$ and $\hat{y}_n(x,t)$ denote the model predictions. Given the initial conditions, the model (7.11) can generate a prediction $\hat{y}(t)$ at any time t. Combined with (7.12), this low-order model can reproduce the spatio-temporal output over the whole time/space domain.

7.3 Nonlinear PCA Based Spatio-Temporal Modeling in Neural System

7.3.1 *Neural Network for Nonlinear PCA*

The optimization (7.9) can be solved using different mathematical tools, which will lead to various NL-PCA implementations. One popular approach is to use a five-layer feedforward neural network (Kramer, 1991). As shown in Figure 7.2, the first and fifth layers are the input and output layers, respectively. The three hidden layers are the projection layer for learning *G*, the bottleneck layer for dynamic modeling, and the reconstruction layer for learning *H*, respectively. The dimension of the bottleneck layer has been significantly reduced. The spatial variable x is added in the reconstruction layer to possess the capability of processing the spatial information.

Note that the nodes of projection and reconstruction layers will be nonlinear activation functions (e.g. sigmoid functions) for modeling arbitrary *G* and *H*. The nodes of input layer, output layer and bottleneck layer can be linear functions. The projection and reconstruction functions can be expressed as

$$y(t) = G(y_x(t)) = W_2\Phi_2(W_1 y_x(t) + b_1) + b_2\,, \tag{7.13}$$

$$y_n(x,t) = H(x, y(t)) = W_4\Phi_4(W_3 y(t) + V_3 x + b_3) + b_4\,, \tag{7.14}$$

where W_i and b_i denote the weight and bias between the ith-layer to the $(i+1)^{\text{th}}$-layer ($i = 1,...,4$), V_3 denotes the weight of the spatial variable x between the 3th-layer and the 4th-layer, and Φ_i denotes the activation function of the ith-layer ($i = 2,4$).

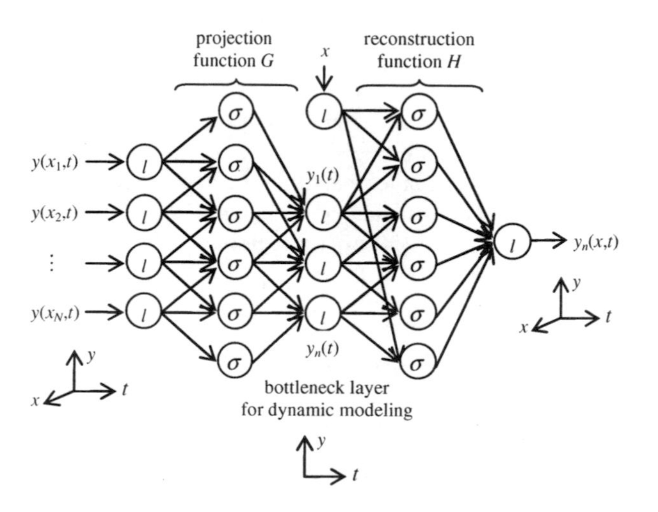

Fig. 7.2 NL-PCA network
(σ indicates sigmoid nodes, l indicates linear nodes)

Nonlinear PCA network design

The network size is critical to the nonlinear dimension reduction. The number of nodes in the projection and reconstruction layers is related to the complexity of nonlinear functions G and H. The number of nodes in the bottleneck layer denotes the intrinsic dimension of the time series. Once the network structure is determined, the maximum capability of the projection and the reconstruction is fixed.

In practice, there should be large enough number of nodes for the high accuracy. However, because of finite length of training data available, too many nodes may cause the over-fitting problem. The tradeoff between the training accuracy and the risk of the overfitting can be managed using some techniques such as the cross-validation (Kramer, 1991; Wilson, Irwin & Lightbody, 1999) and regularization (Hsieh, 2001; Wilson, Irwin & Lightbody, 1999).

The intrinsic dimension of the time series can be determined by the sequential NL-PCA (Kramer, 1991) or hierarchical NL-PCA (Saegusa, Sakano & Hashimoto, 2004) in advance. The linear PCA can also give an initial guess. In this study, we define a measure of the proportion of variance in the data explained by the reconstruction as in Webb (1996)

$$\eta = 1 - \sum_{i=1}^{N}\sum_{t=1}^{L}(y(x_i,t) - y_n(x,t))^2 / \sum_{i=1}^{N}\sum_{t=1}^{L}(y(x_i,t) - \overline{y}(x_i))^2 , \tag{7.15}$$

where $\overline{y}(x_i) = < y(x_i,t) >$ is the mean of the data at the location x_i. Note that the criterion (7.5) in linear PCA is a special case of (7.15), but this measure can help to determine the intrinsic dimension in nonlinear case.

Nonlinear PCA network training

This multi-layer network can be trained by any appropriate algorithm such as backpropagation (Ham & Kostanic, 2001) to minimize the reconstruction error (7.9). The training of the multi-layer network involves a nonlinear optimization where the solution of the weights may be sensitive to the initial conditions. Recently, Hinton & Salakhutdinov (2006) have proposed an efficient approach to solve this problem.

7.3.2 Neural Network for Dynamic Modeling

In our studies, the model (7.10) is assumed to be the following form

$$y(t) = Ay(t-1) + \overline{F}(y(t-1)) + Bu(t-1) + e(t) , \tag{7.16}$$

where the matrices $A \in \mathbb{R}^{n\times n}$ and $B \in \mathbb{R}^{n\times m}$ denote the linear part, and $\overline{F}: \mathbb{R}^n \to \mathbb{R}^n$ denotes the nonlinear part. It can be easily seen from (3.31) that the system in the simulation studies can give a linear and nonlinear separated model when using linear PCA for the dimension reduction. This model can still be used in the NL-PCA based modeling since it may reduce the complexity of nonlinear function F in (7.10). Moreover it is an affine model in the input and then many nonlinear control algorithms developed for affine ODE models can be easily extended to the nonlinear DPS.

In the identification procedure, $\overline{F}$ is approximated as a RBF network, then the model (7.16) can be rewritten as

$$y(t) = Ay(t-1) + WK(y(t-1)) + Bu(t-1) + e(t) , \tag{7.17}$$

where $W = [W_1,...,W_l] \in \mathbb{R}^{n\times l}$ denotes the weight, $K(\cdot) = [K_1(\cdot),...,K_l(\cdot)]^T : \mathbb{R}^n \to \mathbb{R}^l$ denotes the radial basis function, and l is the number of neurons. The radial basis function is often selected as Gaussian kernel $K_i(y) = \exp\{-(y-c_i)^T \Sigma_i^{-1}(y-c_i)/2\}$, ($i = 1,...,l$) with proper center vector $c_i \in \mathbb{R}^n$ and norm matrix $\Sigma_i \in \mathbb{R}^{n\times n}$.

Different algorithms exist for training the hybrid RBF network. Most of them determine the parameters c_i and Σ_i first, and subsequently unknown parameters A, B and W can be estimated by the recursive least-squares method (Nelles, 2001).

Note that the center c_i is often selected using clustering techniques or placed on a proper uniform grid within the relevant region of the training data. The norm matrix Σ_i is often chosen to be diagonal and contains the variance of the training data. The number of neurons l should be carefully determined using such as cross-validation or regularization technique to avoid the possible overfitting problem.

7.4 Simulation and Experiment

7.4.1 Catalytic Rod

Consider the catalytic rod given in Sections 1.1.2 and 3.6.1. In the simulation, assume the process noise $d(x,t)$ in (3.31) is zero. The temporal input $u_i(t)$=1.1+(4+2*rand*) exp(-*i*/5)sin(*t*/14+*rand*)-0.4exp(-*i*/20)sin(*t*/2+2 *rand*) (*i*=1,…,4) is used, where *rand* is a uniform distributed random function on [0, 1]. Twenty sensors uniformly distributed in the space are used for output measurements. The sampling interval Δt is 0.01 and the simulation time is 7.5. A noise-free data set of 750 data is collected from the system quation. The noise uniformly distributed on [-0.08(max(*y*)-min(*y*)), 0.08(max(*y*)-min(*y*))] is added to the noise-free data to obtain the noisy output. The measured output $y(x,t)$ is shown in Figure 7.3. The first 500 data is used as the training data with the first 250 data as the estimation data and the next 250 data as the validation data. The validation data is used to monitor the training process and determine the NL-PCA network and RBF model size. The remaining 250 data is the testing data.

(1) Learning of nonlinear projection and reconstruction

A NL-PCA network is designed to project the spatio-temporal output $\{y(x_i,t)\}_{i=1,t=1}^{N,L}$ into the low-dimensional temporal data $\{y(t)\}_{t=1}^{L}$. Using the cross-validation method the dimension of the time series is selected as $n=2$ and the number of nodes in the projection and reconstruction layers is set to 40. The reconstruction error $e(x,t)=y(x,t)-y_n(x,t)$ of the NL-PCA network over the estimation, validation and testing data is shown in Figure 7.4. The reduced-dimensional time series $y_1(t)$ and $y_2(t)$ (dashed line) are computed for training and testing the dynamic neural model as shown in Figure 7.5 and Figure 7.6, respectively.

(2) Nonlinear dynamic modeling of time series $y(t)$

To establish the dynamical relationship between $y(t)$ and $u(t)$, a NL-PCA based RBF network (NL-PCA-RBF) model with five neurons is used according to the cross-validation method. The model predictions $\hat{y}_1(t)$ and $\hat{y}_2(t)$ (solid line) are compared with $y_1(t)$ and $y_2(t)$ over the training and testing data as shown in Figure 7.5 and Figure 7.6, respectively. The comparison shows that the hybrid RBF model performs satisfactorily.

(3) Nonlinear time/space reconstruction for $\hat{y}_n(x,t)$

Using the nonlinear time/space reconstruction, the spatio-temporal output $\hat{y}_n(x,t)$ is reproduced from the output data of the NL-PCA-RBF model. As shown in Figure 7.7 the prediction error $e(x,t) = y(x,t) - \hat{y}_n(x,t)$ over the whole data set demonstrates that NL-PCA-RBF model can reproduce the spatio-temporal dynamics of the original system very well.

Comparison with linear PCA based modeling

A linear PCA based RBF (PCA-RBF) model is developed for comparison, where the linear PCA is used for the dimension reduction ($n=2$) and the same RBF network for the modeling according to the cross-validation method. The spatial normalized absolute error $SNAE(t) = \int |e(x,t)| dx / \int dx$ is shown in Figure 7.8. Two indexes: η in (7.15) and root of mean squared error $RMSE = (\int \sum e(x,t)^2 dx / \int dx \sum \Delta t)^{1/2}$ over the estimation, validation and testing data are compared in Table 7.1. It is noticed that the NL-PCA-RBF model has a better performance than the PCA-RBF model. The final modeling error consists of two parts: the dimension reduction error and the RBF model error. Obviously, NL-PCA can preserve more variance and achieve less reconstruction error than PCA. Thus, NL-PCA-RBF model can have a better performance because the RBF model is actually a universal approximator.

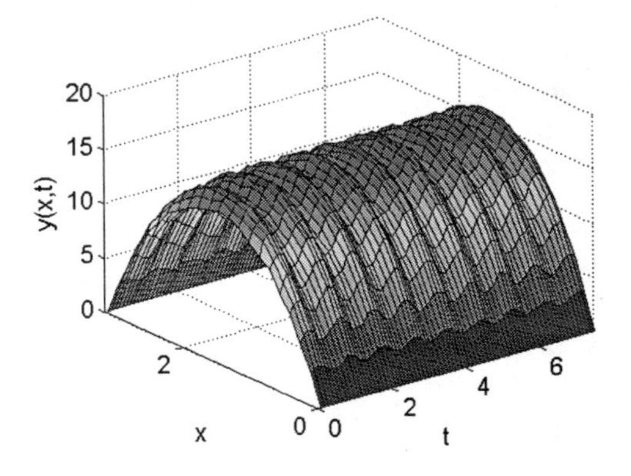

Fig. 7.3 Catalytic rod: Measured output for neural modeling

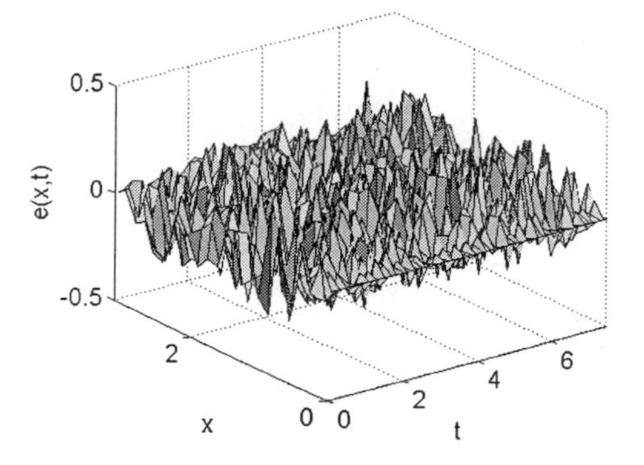

Fig. 7.4 Catalytic rod: NL-PCA reconstruction error

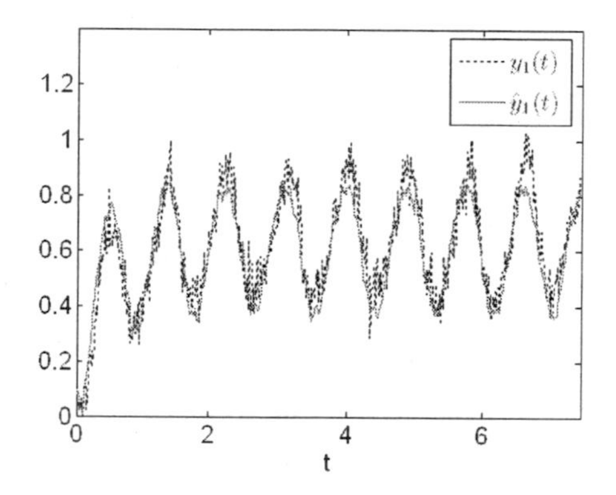

Fig. 7.5 Catalytic rod: NL-PCA-RBF model prediction - $\hat{y}_1(t)$

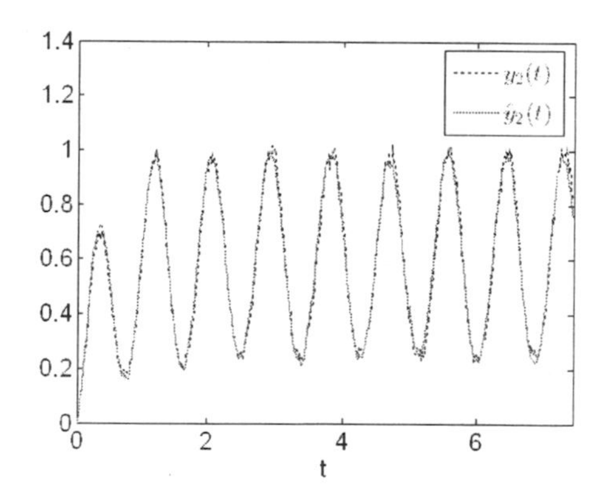

Fig. 7.6 Catalytic rod: NL-PCA-RBF model prediction - $\hat{y}_2(t)$

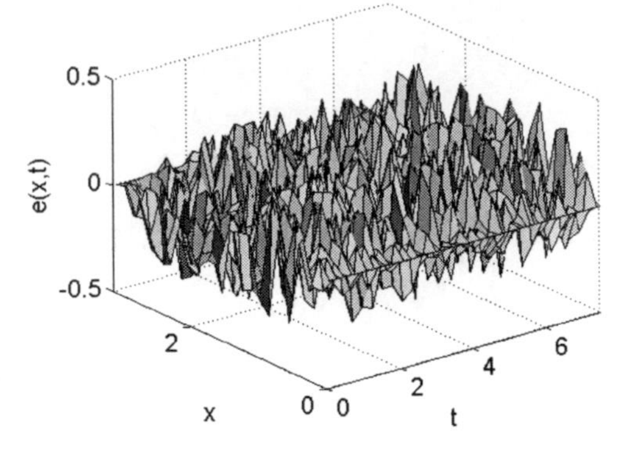

Fig. 7.7 Catalytic rod: NL-PCA-RBF model prediction error

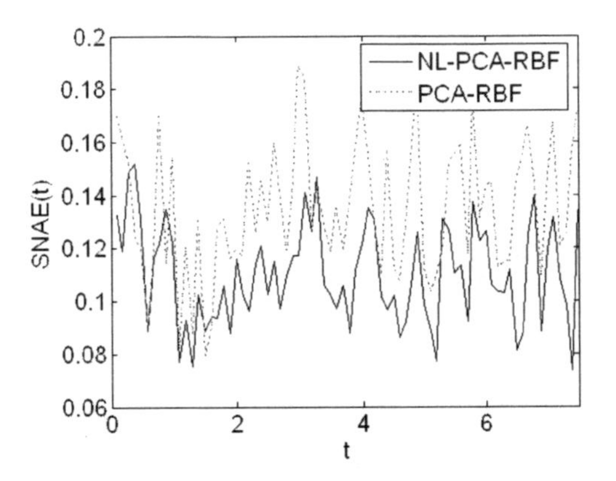

Fig. 7.8 Catalytic rod: *SNAE*(*t*) of NL-PCA-RBF and PCA-RBF models

Table 7.1 Catalytic rod: Comparison of PCA and NL-PCA for modeling

		PCA	NL-PCA	PCA-RBF model	NL-PCA-RBF model
η	estimation	0.9427	0.9566	0.9326	0.9462
	validation	0.9273	0.9458	0.9119	0.9334
	testing	0.9319	0.9508	0.9158	0.9377
RMSE	estimation	0.1407	0.1225	0.1527	0.1364
	validation	0.1417	0.1224	0.156	0.1357
	testing	0.1436	0.1221	0.1597	0.1374

7.4.2 Snap Curing Oven

Consider the snap curing oven (Figure 1.1 and Figure 3.11) provided in Sections 1.1.1 and 3.6.2. In the experiment, a total of 2100 measurements are collected with a sampling interval $\Delta t = 10$ seconds. Among them, 1950 measurements from sensors (s1-s16) are used to train the model, where the first 1800 data is the estimation data and the next 150 data is the validation data. The validation data is used to monitor the training process and determine the NL-PCA network and RBF model size. The last 150 measurements from sensors (s1-s16) are chosen to test the model.

In the NL-PCA-RBF modeling, the number of nodes in the projection, bottleneck, and reconstruction layers in NL-PCA network is 10, 2 and 15 respectively according to the cross-validation method. That means the dimension of the states will be $n = 2$. The model performance at the sensors s1 and s2 over the training and testing data are shown in Figure 7.9 and Figure 7.10 respectively. The predicted temperature distribution of the oven at t=10000s is provided in Figure 7.11.

For comparison, the PCA-RBF model is developed with linear PCA for the dimension reduction ($n = 5$) and the same RBF network structure for the modeling. The performance index *TNAE*(x) (Temporal normalized absolute error, $TNAE(x) = \sum |e(x,t)| / \sum \Delta t$) over the whole data set in Table 7.2 shows that NL-PCA-RBF model is better than PCA-RBF model even with more number of states.

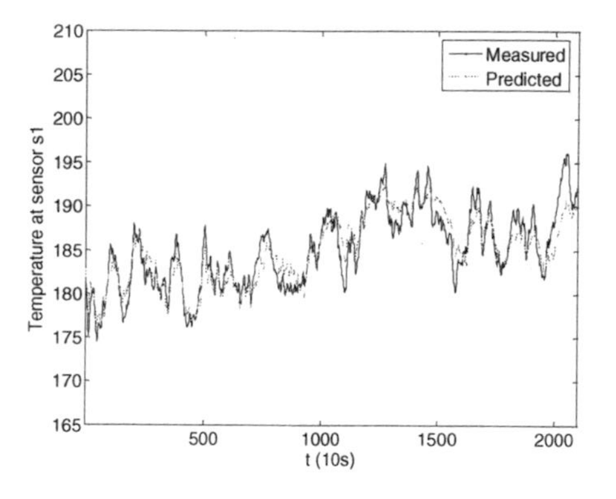

Fig. 7.9 Snap curing oven: Performance of NL-PCA-RBF model at sensor s1

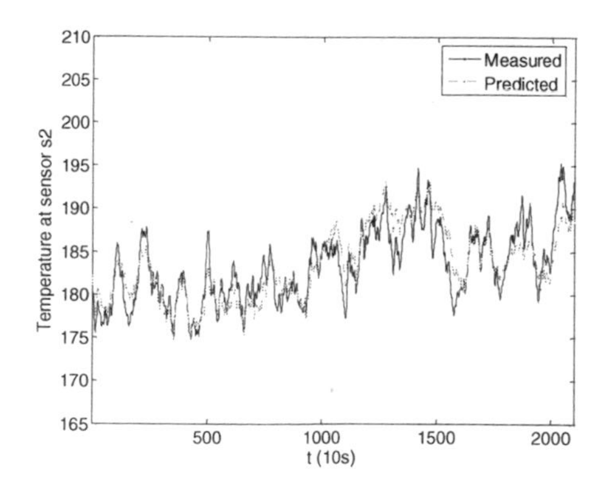

Fig. 7.10 Snap curing oven: Performance of NL-PCA-RBF model at sensor s2

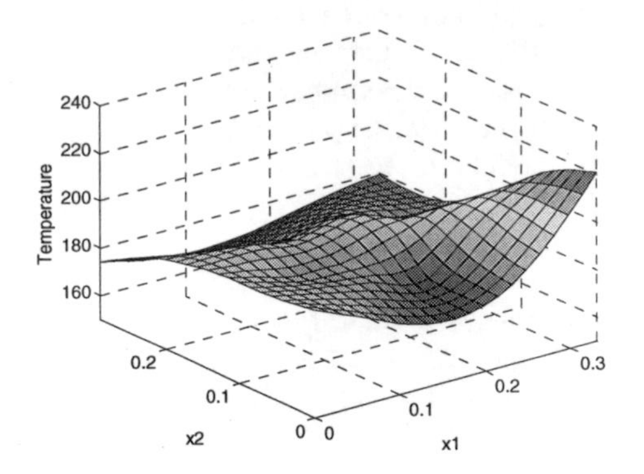

Fig. 7.11 Snap curing oven: Predicted temperature distribution of NL-PCA-RBF model at t=10000s

Table 7.2 Snap curing oven: $TNAE(x)$ of PCA-RBF and NL-PCA-RBF models

	s1	s2	s3	s4	s5	s6	s7	s8
PCA-RBF	1.69	1.85	1.68	1.63	1.65	1.68	1.66	1.61
NL-PCA-RBF	1.41	1.64	1.48	1.51	1.53	1.52	1.52	1.53
	s9	s10	s11	s12	s13	s14	s15	s16
PCA-RBF	1.6	1.9	1.91	1.65	1.78	1.78	1.61	1.87
NL-PCA-RBF	1.69	1.67	1.72	1.66	1.71	1.55	1.63	1.64

Remark 7.1: Performance comparison for modeling of snap curing oven

Different kind of model may be suitable for different system and it is usually selected by the modeling performance. The normalized absolute error $NAE = \int \sum | e(x,t) | dx / \int dx \sum \Delta t$ over the whole data set for all models of snap curing oven (see Chapter 3 - Chapter 7) is shown in Table 7.3.

Table 7.3 Snap curing oven: accuracy comparison for all models

Models	NAE
2nd-order Volterra	0.9225
KL-Wiener	0.99125
3-channel Hammerstein	1.19625
NL-PCA-RBF	1.588125
KL-Hammerstein	1.765

The 2nd-order Volterra model achieves the best performance. The accuracy of KL-Wiener model is similar to 2nd-order Volterra model. Both Volterra and Wiener models are more suitable for this process than other models. The 3-channel Hammerstein model is better than KL-Hammerstein model because of more channels added. Due to the special structure of the oven, the dominant heat conduction is easier to model than the convection and radiation effect. The process is not complex enough to show the advantage of NL-PCA-RBF.

Remark 7.2: ***Comparison of computation time for modeling of snap curing oven***

As shown in Table 7.4, the computation time is also compared for all modeling approaches. Because the time/space separation is often performed individually before the model estimation, the total modeling time can be decomposed into time/space separation time and model estimation time. For the time/space separation, the NL-PCA method for NL-PCA-RBF modeling requires the shortest time. The KL method used in other modeling will take more time than the NL-PCA because of the eigenvalue decomposition of a large matrix in (3.14). After the time/space separation, the NL-PCA-RBF model estimation requires the shortest time. The 3-channel Hammerstein modeling will take more time than KL-Hammerstein modeling because of the multi-channel model estimation. It will take almost same time to estimate other models. Overall, the NL-PCA-RBF modeling is fastest among all the methods used.

Table 7.4 Snap curing oven: computation comparison for all modeling approaches (seconds)

Modeling approaches	T/S separation time	Model estimation time	Total time
2nd-order Volterra	147	42	189
KL-Wiener	147	43	190
3-channel Hammerstein	147	52	199
NL-PCA-RBF	97	3	100
KL-Hammerstein	147	42	189

7.5 Summary

A nonlinear dimension reduction based neural modeling approach is presented for the distributed parameter system. The nonlinear PCA is used for dimension reduction and reconstruction. The traditional RBF network is used for the dynamic modeling. The nonlinear PCA has a more powerful capability of dimension reduction than the linear PCA for nonlinear systems. Thus, the presented approach can work better for nonlinear distributed parameter systems. The simulation on the catalytic rod and the experiment on the snap curing oven have demonstrated the effectiveness of the presented method.

References

[1] Aggelogiannaki, E., Sarimveis, H.: Nonlinear model predictive control for distributed parameter systems using data driven artificial neural network models. Computers and Chemical Engineering 32(6), 1225–1237 (2008)

[2] Armaou, A., Christofides, P.D.: Dynamic optimization of dissipative PDE systems using nonlinear order reduction. Chemical Engineering Science 57(24), 5083–5114 (2002)

[3] Baker, J., Christofides, P.D.: Finite-dimensional approximation and control of non-linear parabolic PDE systems. International Journal of Control 73(5), 439–456 (2000)

[4] Dong, D., McAvoy, T.J.: Nonlinear principal component analysis-based on principal curves and neural networks. Computers and Chemical Engineering 20(1), 65–78 (1996)

[5] Ham, F.M., Kostanic, I.: Principles of neurocomputing for science and engineering. McGraw-Hill, New York (2001)

[6] Hinton, G.E., Salakhutdinov, R.R.: Reducing the dimensionality of data with neural networks. Science 313(5786), 504–507 (2006)

[7] Holmes, P., Lumley, J.L., Berkooz, G.: Turbulence, coherent structures, dynamical systems, and symmetry. Cambridge University Press, New York (1996)

[8] Hoo, K.A., Zheng, D.: Low-order control-relevant models for a class of distributed parameter systems. Chemical Engineering Science 56(23), 6683–6710 (2001)

[9] Hsieh, W.W.: Nonlinear principal component analysis by neural networks. Tellus Series A - Dynamic Meteorology and Oceanography 53(5), 599–615 (2001)

[10] Kirby, M., Miranda, R.: The nonlinear reduction of high-dimensional dynamical systems via neural networks. Physical Review Letter 72(12), 1822–1825 (1994)

[11] Kramer, M.A.: Nonlinear principal component analysis using autoassociative neural networks. AIChE Journal 37(2), 233–243 (1991)

[12] Leontaritis, I.J., Billings, S.A.: Input-output parametric models for non-linear systems - Part I: Deterministic non-linear systems. International Journal of Control 41(2), 303–328 (1985)

[13] Malthouse, E.C.: Limitations of nonlinear PCA as performed with generic neural networks. IEEE Transactions on Neural Networks 9(1), 165–173 (1998)

[14] Nelles, O.: Nonlinear system identification: From classical approaches to neural networks and fuzzy models. Springer, Berlin (2001)

[15] Newman, A.J.: Model reduction via the Karhunen-Loève expansion part I: An exposition. Technical Report T.R.96-32, University of Maryland, College Park, Maryland (1996a)

[16] Newman, A.J.: Model reduction via the Karhunen-Loève expansion part II: Some elementary examples. Technical Report T.R.96-33, University of Maryland, College Park, Maryland (1996b)

[17] Park, H.M., Cho, D.H.: Low dimensional modeling of flow reactors. International Journal of Heat and Mass Transfer 39(16), 3311–3323 (1996a)

[18] Park, H.M., Cho, D.H.: The use of the Karhunen-Loève decomposition for the modeling of distributed parameter systems. Chemical Engineering Science 51(1), 81–98 (1996b)

[19] Qi, C.K., Li, H.-X.: Hybrid Karhunen-Loève/neural modeling for a class of distributed parameter systems. International Journal of Intelligent Systems Technologies and Applications 4(1-2), 141–160 (2008a)
[20] Saegusa, R., Sakano, H., Hashimoto, S.: Nonlinear principal component analysis to preserve the order of principal components. Neurocomputing 61, 57–70 (2004)
[21] Sahan, R.A., Koc-Sahan, N., Albin, D.C., Liakopoulos, A.: Artificial neural network-based modeling and intelligent control of transitional flows. In: Proceeding of the 1997 IEEE International Conference on Control Applications, Hartford, CT, pp. 359–364 (1997)
[22] Schölkopf, B., Smola, A., Muller, K.-R.: Nonlinear component analysis as a kernel eigenvalue problem. Neural Computation 10(5), 1299–1319 (1998)
[23] Sjöberg, J., Zhang, Q., Ljung, L., Benveniste, A., Delyon, B., Glorennec, P., Hjalmarsson, H., Juditsky, A.: Nonlinear black-box modeling in system identification: A unified approach. Automatica 31(12), 1691–1724 (1995)
[24] Smaoui, N.: Linear versus nonlinear dimensionality reduction of high-dimensional dynamical systems. SIAM Journal on Scientific Computing 25(6), 2107–2125 (2004)
[25] Smaoui, N., Al-Enezi, S.: Modelling the dynamics of nonlinear partial differential equations using neural networks. Journal of Computational and Applied Mathematics 170(1), 27–58 (2004)
[26] Webb, A.R.: An approach to non-linear principal components analysis using radially symmetric kernel functions. Journal Statistics and Computing 6(2), 159–168 (1996)
[27] Wilson, D.J.H., Irwin, G.W., Lightbody, G.: RBF principal manifolds for process monitoring. IEEE Transactions on Neural Networks 10(6), 1424–1434 (1999)
[28] Zhou, X.G., Liu, L.H., Dai, Y.C., Yuan, W.K., Hudson, J.L.: Modeling of a fixed-bed reactor using the KL expansion and neural networks. Chemical Engineering Science 51(10), 2179–2188 (1996)

8 Conclusions

Abstract. This chapter summarizes all the methods introduced in the book, and discusses future challenges in this area.

8.1 Conclusions

The studies of the nonlinear DPS become more and more active and important because of advanced technological needs in the industry. The nonlinear DPS considered in this book includes, but is not limited to, a typical class of thermal process in the IC packaging and chemical industry. In general, the modeling is required for many applications such as prediction, control and optimization. However, besides the complex natures of the system including the nonlinear time-space coupling dynamics, unknown parameters and structure uncertainties will make the modeling more difficult and challenging.

After an overview of DPS modeling, the book focuses on the model identification of the unknown nonlinear DPS. The existing DPS identification methods may have some limitations for modeling the nonlinear DPS. For example, the linear kernel model can not model nonlinear dynamics, the linear dimension reduction may not be very efficient for the nonlinear system, and the very complex neural network model may result in a difficult control design. In lumped parameter processes, the popularly used simple structure models: Wiener, Hammerstein and Volterra, have only temporal nature without spatial variables. On the other hand, in the machine learning field, NL-PCA is a nonlinear dimension reduction method, which can achieve a lower order and more accurate model than the KL method for a nonlinear problem. Thus to overcome these limitations in the DPS identification, it is necessary to develop new modeling approaches with the help of different modeling and learning techniques for the nonlinear DPS.

This book is to extend the traditional Wiener/Hammerstein/Volterra/neural modeling to the nonlinear DPS with help of spatio-temporal separation. First, novel concepts of the spatio-temporal Wiener/Hammerstein/Volterra model are constructed, upon which some data-based spatio-temporal modeling approaches are developed for the nonlinear DPS, and applied on typical thermal processes in IC packaging and chemical industry. Based on the presented solutions in this book, the following conclusions can be made.

(1) A spatio-temporal Wiener modeling approach is presented in Chapter 3. For modeling the nonlinear DPS, a spatio-temporal Wiener model (a linear DPS followed by a static nonlinearity) is constructed. After the time/space separation, it can be represented by the traditional Wiener system with a set of spatial basis functions.

H.-X. Li and C. Qi: Spatio-Temporal Modeling of Nonlinear DPS, ISCA 50, pp. 167–171.
springerlink.com

To achieve a low-order model, the KL method is used for the time/space separation and dimension reduction. Finally, unknown parameters of the Wiener system are estimated with the least-squares estimation and the instrumental variables method to achieve consistent estimation under noisy measurements.

(2) A spatio-temporal Hammerstein modeling approach is presented in Chapter 4. To model the nonlinear DPS, a spatio-temporal Hammerstein model (a static nonlinearity followed by a linear DPS) is constructed. After the time/space separation, it can be represented by the traditional Hammerstein system with a set of spatial basis functions. To achieve a low-order model, the KL method is used for the time/space separation and dimension reduction. Then a compact Hammerstein model structure is determined by the orthogonal forward regression, and their unknown parameters are estimated with the least-squares method and the singular value decomposition.

(3) A multi-channel spatio-temporal Hammerstein modeling approach is presented in Chapter 5. As a special case of the model described in Chapter 4, a spatio-temporal Hammerstein model is constructed with a static nonlinearity followed by a linear spatio-temporal kernel. When the model structure is matched with the system, a basic single-channel identification algorithm with the algorithm used in the Chapter 4 can work well. When there is unmodeled dynamics, a multi-channel modeling framework may be needed to achieve a better performance, partially because more channels used can attract more information from the process and also increase the model complexity to match the more complicated process. The modeling convergence can be guaranteed under noisy measurements.

(4) A spatio-temporal Volterra modeling approach is presented in Chapter 6. The traditional Green's function is widely used for the DPS modeling. However, it consists of a single spatio-temporal kernel and is only a linear approximation for a nonlinear system. To model the nonlinear DPS, a spatio-temporal Volterra model is presented with a series of spatio-temporal kernels. It can be considered as a nonlinear generalization of Green's function or a spatial extension of the traditional Volterra model. To obtain a low-order model, the KL method is used for the time/space separation and dimension reduction. Then the model can be estimated with a least-squares algorithm with the convergence guaranteed under noisy measurements.

(5) A NL-PCA based neural modeling approach is presented in Chapter 7. The KL based neural model is widely used for modeling DPS. However, the KL method is a linear dimension reduction which is only a linear approximation for a nonlinear system. To get a lower-order or more accurate solution, a NL-PCA based neural modeling framework is proposed. One NL-PCA network is trained for the nonlinear dimension reduction and the nonlinear time/space reconstruction. The other neural model is to learn the system dynamics with a linear/nonlinear separated model structure. With the powerful capability of dimension reduction and the intelligent

learning, this approach can model the nonlinear complex DPS with much more complexity.

The effectiveness of the presented modeling approaches are verified on some typical thermal processes. Of course, they can also be applicable for other industrial distributed parameter systems.

The presented modeling approaches have the following advantages.

- Little process knowledge is required, as all these methods are data-based approaches. Thus they are more flexible because many real-world applications are in unknown environment.
- Low-dimensional states are obtained. Because of the KL method and the NL-PCA used for dimension reduction, low-dimensional models are usually obtained which are computationally fast for the real-time implementation.
- Nonlinear nature is represented in simple structures. Since the Wiener, Hammerstein, Volterra models are used, the nonlinear models developed are usually have relatively simple structures, which will help the potential applications.

Because almost no process knowledge is required in the modeling, the experiment may need many sensors for collecting sufficient spatio-temporal dynamics information. The number of sensors used should be greater than the number of necessary spatial basis functions. Otherwise, a satisfactory model may not be available. When only a few sensors can be installed, if a nominal model is known in advance and only some nonlinearities are unknown, hybrid modeling approaches, such as, the neural observer spectral method in Section 2.5.4, would be applicable. Even though many sensors may be needed in the modeling, once the model is established, only a few sensors should be sufficient for real-time applications, e.g., prediction and control.

Further investigations on the control design using these models are necessary. There are some challenges which have not been well studied in system identification of the DPS.

(1) Due to the complexity of the process, different modeling methods work in different conditions. Most of the existing methods can only work in the proximity of the operating conditions due to the nonlinearity of the process. Obtaining a nonlinear model which can work very well in a wide range of the working condition is a challenging problem. The identification based on the multi-model methodology is a promising approach. Generally speaking, more input-output data persistently excited at a wide range of working conditions are needed.

(2) The placement (number and locations) of sensors and actuators are very important for the identification of DPS. The modeling accuracy may depend significantly on the choice of sensor number and locations. The actuator number and locations have an important effect on the capability of the persistent excitation in the modeling experiment. Generally, the locations and the number of actuators and sensors are not necessarily dictated by physical considerations or by intuition and, therefore, some systematic approaches should still be developed in order to reduce the cost of instrumentation and to increase the efficiency of identification.

(3) Control application is one of the major purposes for the DPS identification. A proper integration of both system identification and control for an optimal performance is very important and needs much more effort. Though there are many studies reported for LPS (e.g., Hjalmarsson, 2005; Barenthin & Hjalmarsson, 2008; Hildebrand & Solari, 2007), only a few studies have been presented for DPS (Helmicki, Jacobson, & Nett, 1990, 1992; Ding, Johansson & Gustafsson, 2009). A key problem in identification for control (e.g., robust control) is how to measure the error between the model and the unknown system from the data. This problem becomes more difficult in the DPS identification.

(4) It is noticed that current identification techniques in DPS belong to the open-loop identification. Closed-loop experiments could be advantageous in certain situations (e.g., unstable systems), which may be related to the problem of identification for control since a popular method of the iterative identification and control often needs a closed-loop operation. The extension of the closed-loop identification popularly used in LPS (Forssell & Ljung, 1999; Zhu & Butoyi, 2002) to DPS needs an intensive investigation.

(5) Optimal experiment design (e.g., selection of input-output variables, input signal design, and selection of working conditions) to obtain informative data is very important for parameter estimation and system identification of DPS. The placement (number and locations) of sensors and actuators are also critical in this topic. It is a difficult problem, particularly in nonlinear, multivariable, or hybrid case (with both continuous and discrete variables). Only a few studies are reported for parameter estimation of DPS (Qureshi, Ng & Goodwin, 1980; Rafajłowicz, 1983). The extension of the results on experiment design extensively studied in LPS identification (e.g., Forssell & Ljung, 2000; Bombois *et al.*, 2006; Pronzato, 2008) to DPS needs an intensive investigation.

References

[1] Barenthin, M., Hjalmarsson, H.: Identification and control: Joint input design and H-infinity state feedback with ellipsoidal parametric uncertainty via LMIs. Automatica 44(2), 543–551 (2008)

[2] Bombois, X., Scorletti, G., Gevers, M., Van den Hof, P.M.J., Hildebrand, R.: Least costly identification experiment for control. Automatica 42(10), 1651–1662 (2006)

[3] Ding, L., Johansson, A., Gustafsson, T.: Application of reduced models for robust control and state estimation of a distributed parameter system. Journal of Process Control 19(3), 539–549 (2009)

[4] Forssell, U., Ljung, L.: Closed-loop identification revisited. Automatica 35(7), 1215–1241 (1999)

[5] Forssell, U., Ljung, L.: Some results on optimal experiment design. Automatica 36(5), 749–756 (2000)

[6] Helmicki, A.J., Jacobson, C.A., Nett, C.N.: Control-oriented modeling and identification of distributed parameter systems. In: Chen, G., et al. (eds.) New Trends and Applications of Distributed Parameter Control Systems - Proceedings of the 1989 IMA Workshop on Control of Distributed Parameter Systems, ch. 10. Marcel-Dekker, New York (1990)
[7] Helmicki, A.J., Jacobson, C.A., Nett, C.N.: Control-oriented modeling of distributed parameter systems. Journal of Dynamic Systems, Measurement, and Control 144(3), 339–346 (1992)
[8] Hildebrand, R., Solari, G.: Identification for control: Optimal input intended to identify a minimum variance controller. Automatica 43(5), 758–767 (2007)
[9] Hjalmarsson, H.: From experiment design to closed-loop control. Automatica 41(3), 393–438 (2005)
[10] Pronzato, L.: Optimal experimental design and some related control problems. Automatica 44(2), 303–325 (2008)
[11] Qureshi, Z.H., Ng, T.S., Goodwin, G.C.: Optimum experimental design for identification of distributed parameter systems. International Journal of Control 31(1), 21–29 (1980)
[12] Rafajłowicz, E.: Optimal experiment design for identification of linear distributed parameter systems: Frequency domain approach. IEEE Transactions on Automatic Control 28(7), 806–808 (1983)
[13] Zhu, Y.C., Butoyi, F.: Case studies on closed-loop identification for MPC. Control Engineering Practice 10(4), 403–417 (2002)

Index

LaVergne, TN USA
06 April 2011
223147LV00007B/7/P